ADVANCES IN
GROUP PROCESSES

Volume 11 • 1994

ADVANCES IN GROUP PROCESSES

Volume Editors: BARRY MARKOVSKY
KAREN HEIMER
JODI O'BRIEN
Department of Sociology
University of Iowa

Series Editor: EDWARD J. LAWLER
Department of Organizational
Behavior
Cornell University

VOLUME 11 • 1994

 JAI PRESS INC.

Greenwich, Connecticut *London, England*

CONTENTS

LIST OF CONTRIBUTORS

Peter J. Burke

Department of Sociology
Washington State University

Thomas J. Fararo

Department of Sociology
University of Pittsburgh

Lee Freese

Department of Sociology
Washington State University

L. N. Gray

Department of Sociology
Washington State University

Karen Heimer

Department of Sociology
University of Iowa

Norman P. Hummon

Department of Sociology
University of Pittsburgh

D. H. Judson

Bureau of Business and Economic Research
University of Nevada, Reno

Edward J. Lawler

Department of Organizational Behavior
Cornell University

Nan Lin

Department of Sociology
Duke University

Hein F. M. Lodewijkx

Department of Social and Organizational
 Psychology
University of Utrecht

Michael J. Lovaglia

Department of Sociology
University of Iowa

Barry Markovsky

Department of Sociology
University of Iowa

Jodi O'Brien

Department of Sociology
University of Iowa

Jacob M. Rabbie

Department of Social and Organizational
  Psychology
University of Utrecht

Cecilia Ridgeway

Department of Sociology
Stanford University

Lynn Smith-Lovin

Department of Sociology
University of Arizona

PREFACE

EDITORIAL POLICY

The purpose of this series is to publish theoretical analyses, reviews, and theory-based empirical papers on group phenomena. The series adopts a broad conception of "group processes." In addition to topics such as status processes, group structure, and decision making, the series considers work on interpersonal behavior in dyads (i.e., the smallest group), individual-group relations, intergroup relations, and social networks. Contributors to the series include not only sociologists but also scholars from other disciplines such as psychology and organizational behavior.

The series is an outlet for papers that are longer, more theoretical, and more integrative than those published by the standard journals. It places a premium on the development of testable theories and on theory-driven research programs. The editors are particularly receptive to work falling into the following categories:

1. Conventional and unconventional theoretical work, from broad meta-theoretical and conceptual analyses to refinements of existing theories and hypotheses. One goal of this series is to advance the field of group processes by promoting theoretical work.

2. Papers that review and integrate programs of research. The current structure of the field often leads to the piecemeal publication of different parts of a program of research. This series offers those engaged in programmatic research on a given topic an opportunity to integrate their published and unpublished work into a single

paper. Review articles that transcend the author's own work are also of considerable interest.

3. Papers that develop and apply social psychological theories and research to macrosociological processes. One premise underlying this series is that relationships between macro- and microsociological processes warrant more systematic and testable theorizing. The series encourages development of the macrosociological implications embedded in social psychological work on groups.

CONTENTS OF VOLUME 11

Each of the eight chapters in this volume makes a theoretical contribution to a subdiscipline of the small groups research area. Two of the chapters address issues traditionally associated with symbolic interactionist social psychology—but in decidedly nontraditional ways. The chapter by Freese and Burke attempts no less than a general, rigorous theory of social interaction. Their approach integrates social systems, resources, situations, persons, identities, symbolic processes, sign processes, goals, and agendas. Through their individual identities and the situations they create, people create social systems. Social systems, through the situations they affect, reciprocally influence the people that make them. This micro-macro reciprocation is also evident in the analysis of affect control theory and expectation states theory conducted by Ridgeway and Smith-Lovin. Drawing parallels and contrasts between these highly formalized theories of interaction, the authors show that culture plays crucial roles for both insofar as infusing local interactions with broader meaning and purpose.

Lovaglia's chapter also employs expectation states theory in building an integrative approach to the interplay between power and status. In addition to the theory of status characteristics and expectation states, Lovaglia draws upon theories of attribution, exchange and emotion in generating several important new hypotheses. The chapter by Rabbie and Lodewijkx also reviews a highly integrative theory. Their Behavioral Interaction Model bridges intra- and intergroup relations, focusing here on intergroup aggression and conflict processes.

Four of the chapters employ or develop social network imagery in creative ways. Markovsky and Lawler develop a new structural conceptualization of group solidarity and illustrate its use in a formulation of emotion-based group formation. Nan Lin's chapter develops his social resource theory to show how resources embedded in social network ties permit actors to extend their reach beyond the limitations of primordial groups. Among the results are collective rules and broader social structures. The chapter by Judson and Gray develops a new theory of group structural effects on decision making. The theory borrows from a physical theory of coherent structure, and the authors illustrate its ability to accommodate traditional group decision-making rules and structures. Finally, Fararo and Hummon's contribution demonstrates the utility of discrete event computer simulations for developing theoretical models. This exciting methodology can be used to build theories in which social dynamics take center stage. The method is demonstrated in models of network exchange processes and the emergence of larger dominance structures from dyadic social interactions.

Barry Markovsky
Karen Heimer
Jodi O'Brien
Volume Co-Editors

PERSONS, IDENTITIES, AND SOCIAL INTERACTION

Lee Freese and Peter J. Burke

ABSTRACT

In this paper we hope to advance the general theory of social interaction. We bring together symbolic and sign processes in the context of an elaborated identity theory that relates social situations of interaction to social systems of interaction. Situations contain persons with their histories and unique identity sets, pursuing simultaneous and sometimes conflicting definitions, goals, and agendas. Systems are organized active resource flows maintained by the resource interactions of persons through their identity sets. Symbolic processes are tied to potential resources and their flows. Sign processes are tied to active resources and their flows. Identity sets, devolving simultaneously from symbol flows in situations and resource flows in social systems, are the linchpins for persons to relate social situations to social systems, and vice versa.

The current understanding of group process depends heavily upon theories that implicitly or explicitly assume symbolic actors or rational agents that behave in the

Advances in Group Processes, Volume 11, pages 1–24.
ISBN: 1-55938-857-9

process. But how much of the organization and structure of human group process depends upon "pre-symbolic" and "pre-rational" activity? In truth, nobody knows. In this paper we theorize about how unsymbolized interactions are processed by persons and become tied to symbolic interactions, and how by means of the processing (which may or may not look to be "rational") persons tie their situations of interpersonal interaction to their larger (social) systems of interaction.

There is no doubt that humans are more than symbolic actors and rational agents—no doubt that "thinking beings" have more to think about than utilities and more to think with than linguistic symbols. Nonverbal message traffic between humans is "overwhelmingly more copious" than is verbal message traffic—or so claims Thomas Sebeok (1991, p. 14) without apparent fear of rebuke. The noted semiotician went on to observe that "human verbal and various nonverbal means of communication are now so thoroughly intermingled that they can be disentangled only by dint of careful scientific analysis" (p. 33). The various nonverbal means of communication include the nonlinguistic sign, where sign is defined roughly as something that stands for something else or, more smoothly, as something with content that has a perceptible impact on at least one sense organ of an interpretant. A message is a sign or string of signs, and the exchange of messages Sebeok calls "an indispensable characteristic of all terrestrial life forms" (1991, p. 22).

The process of signification ought to have substantial bearing upon social psychological theories of the person, human identity structures, and interpersonal interaction. Now, it will be known to the readers of this volume, the underwriter for much of the current theory of these subjects is symbolic interactionism, an analytical framework built on the bedrock of the linguistic symbol. However, if Sebeok's observations are true, then current theory has left much undiscovered—everything about persons, identities, and social interaction that does not depend upon symbols. That limits the contribution symbolic interaction theories can make toward the understanding of group process. We think the limitations are self-imposed and can be overcome.

We suggest how to modify the conventional theory of identity so that it may include the effects of unsymbolized resource interactions. Resource interactions we conceive as similar to symbolic interactions, but they involve the control and manipulation of resources rather than symbols. Unsymbolized resource interactions, we suggest, form the social psychological experience in terms of which persons interpret their nonlinguistic signs in situations of interpersonal interaction. At the same time, following Freese (1988b), we argue that resource interactions are the ties that bind social systems, for we conceive social systems as *constituted of* interconnected resource flows. We seek a basis for linking persons, their symbolic interactions, their unsymbolized resource interactions, and systemic resource flows. The linchpin is a social psychological processing that we describe using a model for identities, now suitably modified, developed by Burke (1991).

In what follows, we begin by reviewing the theory of identity and pointing out shortcomings in that theory that stem from ignoring all but symbolic interaction. In particular we note the need to incorporate signs and sign behavior into the identity process and we suggest the concept of proto-identities to accomplish this. We then further develop the idea of signs as they relate to resources, which we suggest is similar to the way that symbols relate to meanings. We then suggest that identity sets, which incorporate both the traditional concept of identity and the new concept of proto-identity, act as controllers of both meanings (identities) as well as resources (proto-identities). This is managed by making persons, as agents, be the link between identities and social structure. Finally, we discuss the way that all of this fits together so that people interacting in situations manage both systems of symbol flows and systems of resource flows to create groups and larger social structures and systems. We recognize that what follows is an incomplete effort at theory construction, and we offer no evidence. Nevertheless, we do think the understanding of group process can be advanced, if, as we suggest, group process depends on more than symbolic interaction and interpersonal transactions.

PERSONS AND IDENTITIES

We take our cue from some characteristic views of George Herbert Mead, expressed throughout his work and quoted here from *The Philosophy of the Act*: "In an experience within which individual and environment mutually determine each other, the unity of the environment and its constituent objects as well as that of the individual arises out of the activity of the individual" (Mead 1938, p. 374). For us, as for Mead and many others, persons and their activities are the centerpieces to a theory of social interaction. Mead's reference to the unity of the individual was part of his concept of the self. Mead (1934, 1938) argued that the mediation of the self as an object is the necessary means of interaction between the person and the environment, if new objects are to arise in reflective experience.

For Mead the emergence of mind, self, and society were evolutionary adaptations that conferred for the human species a selective advantage—the advantage of being able to consciously reflect on the conditions of immediate experience by using linguistic communication and role taking as functional activities. Such activities were inherently symbolic. In his analysis of the conversation of gestures Mead emphasized signification as important to immediate experience and as the evolutionary precursor of symbolization. But the main currents of symbolic interaction theory that came after Mead, especially with the influence of Blumer (1969), emphasized that linguistic symbols are more important than nonlinguistic signs as indicators of the conditions of human experience and as a basis of social interaction. With symbolization through shared meanings there comes into being human agency and reflection and all that follows from that.

We, however, wish here to address conditions of experience that may not always be matters of reflection and, therefore, may not always figure into the systems of symbolic understandings in terms of which persons act. There are entire classes of such conditions that bear upon social interaction. Anthropologists working within the framework of cultural materialism (e.g., Harris 1977, 1979, 1989) are quick to point out that the physical constraints imposed upon human action and organization do not always figure into symbolic systems. Social organization and social inter-action are affected by feedbacks that cycle from nonsymbolic through symbolic interactions and back. We have, therefore, to conceive ways to integrate nonsymbolic and symbolic interactions, in order to understand how persons behave and social structures function.

We cannot do this using the concept of self as this is generally understood in current theory, because there has to be included the sorts of things from which self is supposed to have evolved—the immediate experiences involving signification that are prior to as well as coterminous with self. Self is generally thought to depend upon the social character of human language[1] and the socially derived meanings by which objects, including self, are formed and identified through social interaction. We shall not use the general concept of self, for the concept of identity offers richer possibilities for our purpose. Identities are just the sorts of things that persons must possess if they are to be both proactive and reactive,[2] two conditions generally accepted as essential. Persons must be proactive to make any difference to inter-personal interaction, and they have to be reactive for interaction to make any difference to them. If persons and their environments mutually determine each other in a dynamical process of interaction, then a theory of microscopic interaction must rest on two things: (1) a theory of the person, as both cause and effect, and (2) the manner in which interaction dynamics are organized by identity dynamics. This orientation would seem to put us foursquare in line with the classical origins of identity theory in social psychology, but we think these origins are somewhat misleading.

There are several classical notions of identity (Weigert 1983). It is the specific inheritance bequeathed by Nelson Foote (1951), and now belonging to conven-tional sociological wisdom, which has enhanced and expanded it, that we wish to highlight. The idea that has evolved from the influence of Foote is to interpret the concept of identity as a relation between the person and social structure. The process of identification, in this view, is thought to proceed by the naming and labeling of persons as belonging to socially structured statuses or roles whose concatenation in situations of interaction define a person's identity. Through a succession of situations of social interaction a person's identity becomes a binding thread that links the person to the groups to which he or she belongs, through establishing membership in a common social category. An identity also is thought to establish the uniqueness of the person, and so the concept does double duty: A person is identified *with* others (in structured positions) while also being individuated or distinguished *from* them. The double duty is necessary, if one is to argue that persons are the unique products of interaction within common social structures and can so experience themselves. And so it has become one of the truisms of symbolic interactionism, and indeed of sociology, that a person has no individuality apart from identity and no identity apart from society.[3] But there is a problem with this.

The development of the theory of identity now becomes unnecessarily self-limiting. The limitation is that identities are symbol-dependent. They can crystallize only because of a normative consensus for social meanings communicated through language. As symbolic interaction theory has it, the principal function of the symbol is to indicate something about an object. An object is a referent for a symbol and a symbol is a significant sign, that is, one having an arbitrary but shared meaning established through social interaction or socially agreed-upon responses (Lindesmith and Strauss 1956). Significant objects, designated by significant signs, in the framework of symbolic interaction theory are defined and differentiated only by means of linguistic symbols. The pattern of circulation in symbolic interactionism is from symbols that define objects and objects that can be defined only symbolically.

Generally left out is the role of signs that are not symbols (i.e., that do not necessarily have shared conventional responses). Sign is a more general concept than symbol. Charles Peirce, who supplied some of the roots of symbolic interactionism as well as of modern semeiotics, held that except for instantaneous consciousness, all thought and knowledge is by signs (Hartshorne and Weiss 1958, p. 390). Peirce did not hold that all signs were symbols in the symbolic interactionist sense of the term. Subsequent to Mead's analysis of the conversation of gestures, in symbolic interaction theory nonlinguistic signs in general have given way to symbols in particular, which are usually understood to be the signs we share through language (broadly conceived).[4] On this has generally stood the theory of identity and of self. The objects of symbolic (reflective) experience are included. The objects of sign (immediate) experience are not—they generally are not even taken as objects, since they are unsymbolized.

However much language may matter for social interaction, it is not all that matters and, if not, then symbols are not all that matter—a result that we ought to accept on other grounds. A great many social animals have complicated forms of social interaction and organization without taking recourse to linguistic symbols. They use nonlinguistic signs.[5] There is no reason to suppose—and certainly no evidence—that the rich symbol systems of humans have caused the degeneration of human knowledge based on nonlinguistic signs, or have eliminated all personal and social implications of that knowledge. To the contrary, verbal and nonverbal signs are intricately comingled in the human repertoire, as (following Sebeok) we noted earlier. How can those implications be explored for the present theoretical purposes?

The only way is to turn a classic notion on its head—specifically, the notion that identities link persons and social structure. We can do this without denying that identities have social sources, at least to a significant degree. Let us affirm that. But rather than taking identities as relations, we propose to reverse this action. *We shall take persons as the link between identities and social structure.* This will have two advantages. First, it will enable the person to be more proactive—a desirable condition if interpersonal interaction is to generate social structure. Second, it will enable us to incorporate the effects of interaction conducted with nonlinguistic signs as well as with linguistic symbols—a condition we hypothesize to be necessary for interpersonal interaction to generate social structure.

It might be objected that too many eggs are scrambled with this proposal to include signs as well as symbols into a conception of identity. Identities are symbolic constructions that emerge from social interaction. What do nonlinguistic signs have to do with this? Unlike the meanings attached to linguistic symbols, the meanings attached to nonlinguistic signs do not require language or social consensus. Therefore, it would seem, they cannot be associated with identities. What is the point of suggesting, as we do, that an identity contains two distinct meaning processes, symbolization and signification?

There are several points. First, to maintain that signification and symbolization are two distinct meaning processes is dogmatic. While the dogma may have had its uses, it also has obscured as much as it has illuminated.[6] As a matter of principle there is no reason that theory cannot be extended to include both processes. The meanings associated with symbols as well as with signs are normally taken to derive from the reactions provoked by a given symbol or sign, so this affords no basis to distinguish the two. Both are indicators for appropriate responses to environmental input. The difference is that the meanings of symbols are taken to be arbitrary social conventions, and the meanings of signs are not. But persons have to process the meanings of both and act according to those meanings, and there is no a priori reason to think that signs and symbols are processed differently (cf. Lindesmith and Strauss 1956).

Second, social primates clearly signal, distinguish, and recognize each other, and can identify themselves in relation to others (Kinzey 1987; Box 1984; de Waal 1989; Goodall 1986; Heltne and Marquardt 1989). They attach meanings to signs, but maybe not to linguistic symbols.[7] Whether or not there is a true identification process here is very muddled, but there clearly is organized social interaction in which identifiable and identified individuals make a big difference. Much if not all of that difference depends on the animals' interpretation of signs. We could whisk away the inconvenient facts of primate social interaction and organization as having no bearing on the case of humans, but this would not be in accord with the science of our day—particularly since humans are primates. Better to acknowledge that we and these nonhuman cousins are doing some of the same things.

So as not to expunge too much convention all at once let us make a semantic concession. Part of the classic notion of self, which we said we would not touch, is the notion of identity, which we now seem to be profaning. If self requires mind and mind requires symbols, how could it be otherwise for identities? Mead argued that without mind animals could recognize sensuous characters but not objects, and the argument has stuck (1938, p. 331). So, for the time being, let us use the term *proto-identity*[8] to designate a person's organization of meanings associated with nonlinguistic signs, or the objects of immediate experience; and let us reserve *identity* for the organization of symbol meanings, or the objects of reflective experience.[9] We may provisionally proceed as if signs and symbols provide different material for interaction with which the person must come to terms, and that persons do so by separately organizing and processing these different sorts of meanings. The concept of proto-identity is adopted to suggest that persons are capable of accumulating sign meanings in which they as persons are implicated, during the course of their experience, into something like an identity even without the linguistic symbols that identities are taken to require. The terminological distinction, artificial and simplistic to our way of thinking, is intended as a temporary heuristic device. Upon eventual analysis of the processing we suppose to occur for signs and symbols, and for what we take the appearance of signs and symbols to mean in the experience of persons, we shall see that the distinction can be dispensed with.

SIGNS, RESOURCES, AND SYMBOLS

What is important about proto-identities, and what do they have to do with persons and their social interaction? While both Peirce and Mead provide some thoughts appurtenant here, even taken together they only provide a start for what we wish to develop.

Signs

Peirce suggested that the whole universe might be entirely composed of signs, and that there was nothing that could not serve as a sign. For him, signs were not things but triadic relations by which the sign connected some object to the thought, action, or experience of an interpretant. His essential idea was that knowledge or awareness of a sign carries with it knowledge of something else. The sign, in effect, being an object of immediate experience, is an indicator of how to respond. Presented with signs, which could consist of almost anything, persons were presented with conditions of thought and action (Peirce 1958; Fisch 1978). Concerned though he was with formal logic, Peirce did not allow a hard and fast distinction between human and nonhuman sign processes (Sebeok 1975).

Neither did Mead. His concept of the gesture, adapted from Wilhelm Wundt, reflects approximately the Peircian notion of sign put to different and more limited uses. Mead locates gestures within social acts—at the beginning of a social act, where a gesture undertaken by an individual calls forth responses by other individuals in a later phase of the act. Conversations of gestures occur because objects of experience are constituted by the meanings they have for the organisms responding to them, and the meaning of a gesture by one organism is the adjustive response made to the gesture by another organism. Thus animals other than humans could converse by means of gestures or signs. What made humans special, in Mead's view, was their use of language, because herein lay the keys to the evolution of mind and self. Language enables significant gestures, namely the conscious calling forth to oneself of the same meaning that one's gesture calls forth to another (Mead 1934). Mead's hard and fast distinction, which served his purpose, was between gesturing in general and symbolizing in particular. Only the latter enabled role taking, which was necessary for the theory he was developing.

Mead's interest was in explaining mind and self. For our purposes, more relevant are his explicit views that gesture is the basic mechanism whereby the social process is conducted (1934, pp. 13–14), that language is a complication of the gesture process, and that language symbols are simply significant gestures (consciously used with shared meanings). As he put it, "Meaning can be described, accounted for, or stated in terms of symbols or language . . . , but language simply lifts out of the social process a situation which is logically or implicitly there already" (1934, p. 79). The social process, as he said, was conducted by gestures or, as we might say, by signification.

This is why we think proto-identities as sign processes are important. It would be myopic to assume, and Mead did not, that all signification in fact was represented by symbolization, or was superseded by it, or that signification was irrelevant to social interaction. If we take Peirce at his word that signs are pandemic and Mead at his word that gesture is the basic mechanism of social process, then the explanation of social interaction must reckon with the signification process of persons—the process whereby they interpret and respond to the meanings of signs. This process is identity-connected for the simple reason that different interpretations of sign configurations, especially incorrect responses to them, may have implications for personal survival; but even those that do not have implications for personal functioning.

They also have social implications. As Mead and many ethologists have observed, cooperative activity can occur without commonly shared meanings being given to signs, though in fact sign meanings are often commonly shared. A collapsing bridge means about the same thing to everybody caught underneath it, and so does a sentry's warning to the bison in a herd. The interpretation of gestures or signs, however, is not necessarily universal nor necessarily a property of consciousness. It is a property of the actual field of immediate experience defined through learning. So being, the configuration of signs in a field of experience, including in an asocial environment, can place persons in attitudes with respect to the field—attitudes determined by the configuration. But what process determines the response?

Mead kept his notion of attitude within a social context, and did not range into private experience to find causes. Proto-identities are partly private affairs, it would seem, since they contain the unsymbolized meanings in the experience of a person.

But, although persons are partly private too, that does not imply their private experience has no measurable public effects. If the private experience of persons affects them, it should affect their social interaction also.

What do persons actually experience in connection with signs? Certainly not the signs as such. On the conventional view, which we do not question, signs convey meanings to persons by providing the persons knowledge of something else upon the appearance of the signs. Signs are markers that structure or configure experience, and the meanings and knowledge provided by variable signs are the substance of the variable experience that sign configurations organize for persons. But, if signs are pandemic, as Peirce held, is there anything pandemic about the meanings or knowledge conveyed notwithstanding the variable information content of different signs? In other words, is there any experience that might be indicated to all persons with the appearance of any sign or sign configuration?

We think there is. Sebeok (1975) hinted at it when he observed that any form of energy transfer can serve as a sign vehicle. In our view, although persons may not always consciously interpret their experience as such, *signs indicate to persons the condition, state, motion, or transfer of resources.*[10] And this is the process that determines persons' responses to the attitudes in which they are placed by their conditions of interaction. Their conditions are configurations of resource flows. Obviously, such a thought needs to be further explained, and we do that in the next section.

Resources

The term resource is unfortunate because—in almost every Western cultural discourse—it tends to have some very limited connotations. The connotations include resources as tangible or intangible, which allows for a decently broad scope; but beyond that the connotations narrow to resources as consumable, negotiable, commodity-like, entity-like, valued, scarce, and available from an environmental background that, as opportunities permit, social actors can transform in the foreground to their benefit. In no science whose theories deal with resources is the concept seriously developed beyond these ordinary cultural, economically derived, connotations. For our purposes we need more than this utilitarian notion.

We proceed at a very abstract level and interpret resources as processes rather than entities. Resource processes are definable by their effects on a system of interaction. Anything that functions as necessary or sufficient to sustain a system of interaction is taken as a resource—whether or not it is valued, scarce, consumable, possessible, negotiable, leveragable, tangible, or even cognizable. This admits almost anything to the category, but with a transposition of the idea. The idea is to focus on *resource functions in process* instead of resources in place (Freese 1988a). Our argument, developed more fully by Freese (1988b), is that *connected resource flows form the fundamental interaction process of any social system, out of which and because of which system structures are constituted.* This includes but goes beyond the conventional idea of resources as zero-sum desiderata available for differential allocation among competing interests. It permits possessed entities of any sort, valuable or not, to be counted among the resources insofar as they behave in transfer processes that function to sustain interaction. Also counted among resources are various assorted conditions that are not entities at all—for example, conditions of sequencing, or of structuring, or of sentiment, or of opportunity.[11]

There is no requirement in our view that persons be consciously aware of, or have any utilitarian interest in, or have any symbolic knowledge of how to utilize, the resource flows that are functioning to sustain them and their systems of interaction. In this view, resources are not supplied from an environment in the background. *Rather, resource flows are the environment—the immediate environment in terms of which persons functionally connect to their social systems.* Our focus is not on what persons and collectivities *consume* as resources, but on what they do that is resourceful. Something is resourceful if it sustains or enhances the interaction process and the persons connected with it. If it does not, it is not.

This generalized concept of resource enables us to provide the first of several theoretical axioms that link identities to social structures. Exchange and rational choice theory can do this only insofar as utilitarian value is at issue. Symbolic interaction theory can do this only insofar as symbolic processes are at issue. But, we think there is more at issue than utilitarian value or symbolic processes. If we interpret resources as functions that sustain the social interaction process, we may connect this idea with signification, which is where this discussion began, by assuming this: *To every configuration of signs there corresponds a set of active resource transfers,* which is to say that persons perceive relevant resource transfers through their signs.

That, we suggest, is what signs mean to the persons (or other organisms) who are interpreting them. For Peirce, almost anything could be a sign; for us, almost anything can be a resource. The connection is that signs are indicators of resources actively in motion. The actual or imminent transfer of resources is the experience that accompanies the appearance of sign configurations. The experience is both input to and (through learning) stored in proto-identities. When persons interact they refer to it.

Before moving on to the concept of symbol, we take a slight digression to make this idea of resources and resource flows a little more concrete. To that end, imagine a steel mill and the kinds of resource flows in and through it that define it. Imagine the flow of iron ore to the mill; the flow of coal and coke to the mill; the flow of electricity to the mill; the flow of water to the mill; the flow of heat, slag, and contaminated water out of the mill; the flow of steel out of the mill; the flow of people into the mill on a work day and home again in the evening; the flow of equipment, supplies, order forms, computers, pencils, trucks, cranes, and so on through the mill; the flow of purchase orders, money, credit, and debt through the mill; the flow of skills, information, actions, and behavior through the mill; the flow of labor; the flow of organizational activities that enable the flows of physical resources and processes to be managed and utilized; the flow of esteem, respect, and power to various managers and workers.

Note that all of these objects, flows and transformations are initiated and enabled by the actions of persons. Note further that the resources themselves, material or intangible, can have no function whatever until they are in motion, that is, until they are flowing in a connected manner. What we want to suggest from the above example is that *the flows and organized transformations of resources at a very abstract level are the social system, in the sense that they constitute it.* And, it is the resource (and resourceful) interactions of individuals that guide and control those flows.

Symbols

The meanings associated with signs are normally taken to be properties of the context of some situation, in which the meaning of a sign to an interpretant is inferred by the response the sign provokes. The concept of a proto-identity is simply our way of suggesting that persons organize their experiential meanings associated with signs and that these meanings become properties of them as persons. Persons carry with them a collection of sign meanings, learned and accumulated from their experience, which they incorporate and integrate into their personal histories as standards for assessing signs in their immediate experience. If they do this, then sign meanings will have implications similar to the implications of the symbolic meanings associated with the person's identities. Especially they will have implications for interpersonal interaction. This is true if, by hypothesis, the appearance of a configuration of signs "means" to persons the transfer of some resource indicated by the configuration.

What do symbols "mean?" Symbols are normally taken to be significant (i.e., consciously shared) signs. But if signs indicate to persons the active flow of resources, then what is there for symbols to indicate? By attributing to signs this universal function, we shall have to do something likewise for symbols. Let us commit to a very general assumption congruent with the first: *To every configuration of symbols there corresponds a set of potential resource transfers.* Symbols are thus the means by which people transcend the experience of immediate signs and the resource transfers that they signal—the means by which they store information about signs learned, make plans about signs, coordinate and regulate their plans and information, in short, how people develop a culture and transmit it.

Distinguishing active from potential resource transfers is approximately analogous to the distinction in thermodynamics between kinetic and potential energy. We think the distinction can be theoretically fruitful because whatsoever functions as a resource at some point in time might not so function at another point, and similarly things that are not now resources may, with planning, become so in the future. It depends on the meaning to the person that the resource has at various points. For example, during some event of expected long duration, being informed of the time may mean nothing early in the event and may mean something important later. Is that information not a resource in both instances? Yes, but it is not functioning in the same mode. In the first instance, the resource is not implicated in any change relevant to the interaction process. It is a potential resource with no effects. When later that same information changes the interaction process, its potential is then realized. An *active* resource is anything that acts to sustain persons or an interaction process in some system. A *potential* resource is anything that could become an active resource if it is positioned properly at the right place and time, from the point of view of the person. Active resource transfers are the objects of immediate experience, indicated by signs. Potential resource transfers are the objects of reflective experience, indicated by symbols.

That is what we take symbol configurations to mean to a person: some unrealized but potential arrangement of resource transfers in which the person is or may be implicated. For our understanding, symbols have a universal function in behavioral contexts, notwithstanding the idiographic content they might convey. It is to indicate the position of a person with respect to potential changes in resource flows. A symbolic configuration may mean to a person that some present resource function could change or that some missing resource function could be activated, thereby altering the web of social interaction in which the person is implicated. Again, we do not claim that persons necessarily understand, report, or explain their symbolic experience in this way. We claim that *is* the underlying experience by which persons connect symbols and objects.

Why take this view? Because all symbolized objects are potential resources. Symbols enable persons (in collaboration with others) to define, interpret, construct, differentiate—bring into their reflective experience—objects of their immediate experience to which the symbol (or set of them) refers. In referring to objects, symbols may have various connotative or denotative meanings in a semantic attitude, but the objects themselves have functional meanings in a behavioral attitude. *The functional meaning of the semantic content of symbols must derive from the functional meaning of the objects symbolized.* And the functional meaning of the object is the resource flow potential. This is not what symbols themselves symbolize; it is what objects themselves symbolize, so to speak, in the sense that resource potential is the functional meaning that objects have for persons. Thus, although symbols may refer to objects, they must indicate more to persons than that reference value—specifically, they must indicate a particular property of the objects, such as their resource potential—a potential that interaction contexts could realize as active flows. The resource potential of objects is *why* persons differentiate, define, and symbolize the objects they do. There is no incentive to symbolize something that is not potentially resourceful.

IDENTITY SETS

We began by noting that there is more to social interaction and society than is contained in things symbolic. To understand the link between self/identity and social interaction we must supplement the idea of symbols and symbolic interaction by reintroducing the idea of signs. Within conventional identity theory, the concept of identity is used to control the flow of symbolic meanings. Signs are stimuli, the meanings of which reflect the current state or flow of resources, either social or physical. To incorporate the role of signs we now extend the notion of identity to a more generalized conception of an *identity set*. With this term we refer to the entire collection of identities and proto-identities of a person. We have a two-fold purpose in doing this.

One purpose is to suggest that proto-identity processes, which govern the flow of sign meanings (via the flow of resources), can be described with a model that has already been applied to identity processes that govern the flow of symbol meanings. One model for both kinds of process would be desirable. Whether or not persons admit into their reflective experience the conditions of their immediate experience, there must be an intimate connection between the two. The conditions of immediate experience, indicated to persons by signs, are the active resource transfers that sustain interaction. The conditions of reflective experience, indicated to persons by symbols, are potential states and changes in those transfers. Social interaction contains simultaneous flows of resources and symbols—that is, signs of active resource functions and symbols about potential changes in them. Persons, armed with identities and proto-identities, must utilize information about their environments in order to interact symbolically and resourcefully. We suggest a model for this in the next section.

A second purpose, implicit in what was just said, is to relate identity set processes to social interaction. The casual use of the term social interaction has long obscured the possibility that there may be a fundamental dimension of social interaction quite distinct from the interpersonal sense of the term. Current theory takes identity processes to be fundamental to interpersonal interaction—an axiom nobody doubts. But what of proto-identity processes, or the meanings associated with signs? How do they—how does the immediate experience of resource flows—organize social interaction? These matters we address two sections hence.

Identity Sets and Interpersonal Interaction

The modern sociological theory of identity has been constructed almost entirely with reference to the symbolic properties of identities and the normative consensus that identification is presumed to require. Given its roots in the work of Foote (1951) and others of like mind, much of this theoretical development is quite logical. For example, the work of Sheldon Stryker (1980, 1987) and his associates construes self as a structure of identities, and places the concept of role identity front and center. On this view the sources of a person's identity are located in the network of roles, statuses, and norms of the person's subcultures—networks that provide a structure for interpersonal interaction and, therefore, a way of locating, defining, and identifying persons. The identities[12] of a person are taken to be organized symbolic meanings relating a person and his or her social structure. With its emphasis on symbolic processes identity theory has not accorded a serious role to signification.

By now it should be clear that we do. Persons relate themselves to their surroundings partly in terms of the signs their surroundings provide. Often, but not always, persons symbolize this. When they do, these meanings are sorted into persons' various organized identities. But even when they do not, sign meanings still form part of persons' experience and have therefore to be sorted into organized proto-identities. Proto-identities represent the unsymbolized immediate experience of active resource transfers. Identities represent the symbolized, reflective experience of potential resource transfers—with implicit expectations about possible changes in resource interactions in which the person is or would be implicated. Given this line of analysis it is hardly radical now to suggest that, *for every identity, there exists at least one proto-identity with which it is correlated.*

If there is some organization of sign meanings into proto-identities and symbol meanings into identities and some correlation between the two, it may be assumed that the person is able to reorganize his or her various identities, proto-identities, and relations between them in response to input from situations of social interaction. We need not assume much now about the nature of this internal organization of meanings and meaning sets, as this is not pertinent to our present goals. It is more pertinent to assume something about the behavior of the entire conglomerate of sets of identities and proto-identities, whatever their organization, distribution, and dynamic internal interaction. Then we can connect the person to social interaction and structures of social interaction.

We have used the term identity set to include both an identity and proto-identity of a person. Since the constituent meanings of both are properties of the person, we take the person's identity sets also to be properties of the person, ones that are imputed by self to self. We doubt that anything makes much of a difference to persons during interpersonal interaction if it is not connected to their identity sets. Our reason is that *identity sets provide an organization to persons' experience that enables them to assess actual or potential resource transfers that affect them.* Therefore, persons ought to respond to their surroundings and affect their surroundings by way of their identity sets.

Perturbations In Identity Sets

If meanings are organized and stored into identity sets, then the stage is set for persons to be both proactive and reactive. Assume this: *In any situation of interpersonal interaction the interruption of actual or potential resource flows for persons perturbs the meaning systems of their identity sets* (cf. Burke 1991). We have already reckoned that signification carries with it the experience of active resource flows and that the primary function of symbolization, insofar as it connects to signification, is to position the person in a field of potential resource flows with respect to the field of active flows. The immediate assumption, which is hardly bold, implies that persons will not be satisfied with a symbolically defined situation of interaction unless they are satisfied with the situationally relevant resource flows, active or potential. If there is not a satisfactory correspondence their meaning systems are perturbed, which is to say their identity sets are activated. The result is, persons are activated.

The significance of identity sets for interpersonal interaction is that persons use the relevant identity and proto-identity standards as referents for the processing of situational input. Persons have to have some way of sorting the meanings made available from their surroundings, and have to have some impetus to sort them, manage them, and turn them into output. The questions of means, motivation, and mechanism of response we introduce by using a model developed by Burke (1991).

Originally used for the traditional concept of identity, this model describes how a person gauges meaning inputs against a situationally relevant identity already provided as a standard. The model does not describe how an identity standard itself may change because of situational interaction. It assumes that the person compares a relevant identity standard with situational meaning inputs, is motivated to reduce any discrepancy in order to avoid stress, and that the person will engage in behavior, in the situation, designed to restore congruity between the meaning inputs and the identity standard. Thus the model describes how meanings are cycled from the (situational) environment through the person and back again so as to involve the person in a continuous and dynamic matching process involving person and situation. The model does not presume the person acts to restore an identity to some dynamic equilibrium, nor does it presume that the attempted matching between personal identity and situational interaction succeeds.

This model is easily generalized for use with identity sets. It is sufficiently abstract in its outline that it may be interpreted to sign meanings as well as to symbol meanings—thus to proto-identities as well as identities and to what we have called identity sets. Persons process actual or potential resource flows because that is what interaction situations present them with. The processing, according to the Burke comparison model, provides persons with readings to gauge the standards of their identity sets that actual or expected resource flows in the situation have made relevant. *Whatever identity sets are activated provide standards to which the person relates or compares the symbol or sign meanings that the situation provides as input. Dynamic behavior by the person in the situation is a function of that comparison.*

Figure 1 shows the model of the identity process as it has been developed to deal with symbolic interaction. On the input (left) side of the figure are the various reflected appraisals growing out of the interactive situation (labeled "symbol flows"

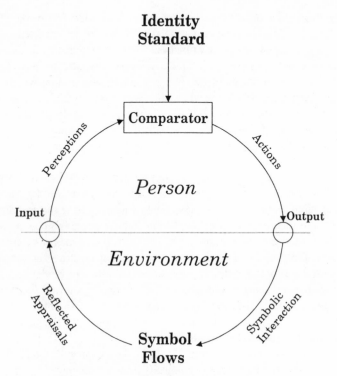

Figure 1. Model of the identity process

in the figure). Perceptions of these appraisals or meanings are sent to a comparator, where they are compared with the meanings contained in the identity standard. To the extent that the perceived meanings do not correspond with the meanings contained in the standard actions, are taken on the output (right) side of the figure. These actions as manifest in the situation are interpreted by self and other in terms of their symbolic value—they are meaningful and contribute to the ongoing symbolic interaction. The design of the actions are to alter the symbol flows in the situation, thus altering the reflected appraisals and perceptions of them so as to bring into correspondence the perceptions and the identity standard. This is a continuously adjustive process of perception, action, and symbolic interaction where the goal of the action is to bring the perceptions into alignment with the identity standard. Action is thus not simply a function of the meanings of the identity standard, nor of the perceptions (meanings) from the situation, but rather is a function of the relation between these.

Since any interaction situation represents a confluence of multiple symbol and sign configurations, and not always the same ones for different persons, continual action and reaction by persons in an interpersonal interaction situation is a matter of course. There is no state of rest. Perturbations in the identity sets of persons are the foundation for continual order-defining and sense making activities, for negotiations, renegotiations, and positioning, for explaining, accounting, and manipulating, for rearranging status and role relationships, and for reconceiving self and others. All this may be interpreted as activity designed to resolve the tensions inherent in active and potential resource flows that multiple interacting persons will differentially experience or interpret. Their identity sets are at issue. That means *they* are at issue.

The Burke comparison model is general enough to describe the processing of information about actual as well as potential resource flows. When applied to entire identity sets, additional matters come into play. If meanings representing actual and potential resource flows are sorted into proto-identities and identities, which themselves become correlated, then the person has a major occupation: It is to resolve the tensions between symbolic and resource interactions when they fail to correspond. This is not just a matter of social cognition or interpersonal negotiation conducted symbolically in the situation itself. The tensions may extend to *unsymbolized* resource flows and symbolized resource potentials, and the social interaction process now becomes more than just interpersonal.

SITUATIONS AND SYSTEMS OF RESOURCE INTERACTIONS

The concept of a social "situation" (or what we at times have called an interpersonal interaction situation) by itself is quite inadequate to provide a context for analyzing the significance of persons in interpersonal interaction. No situated episode of interpersonal interaction is self-contained, nor are repeated episodes governed by the same norms self-contained. All persons, in all situations of interaction, bring with them personal histories, unique identity sets, and common cultural norms that help to contain the interaction and to control it, as well as to individualize it. Moreover, any situation of interpersonal interaction is inordinately complex, because it combines multiple layers of social reality in the form of simultaneous and sometimes conflicting situational definitions, goals, agendas, expectations, and potential rewards and costs. A situation of interaction materializes with a gathering of assorted persons, each bringing assorted interpersonal inputs (signs and symbols). It results in variegated outputs of behavior that is both symbolically meaningful and situationally resourceful (i.e., having impact upon the resource flows in the situation). All of these inputs and outputs may be differently interpreted by different persons. How does interpersonal behavior then become patterned and organized so that microscopic interaction is converted into macroscopic structural effects? Lots of microscopic situational interaction has no discernible macroscopic structural effects at all. Still, we may assume the effects are transmitted there, in interpersonal interaction, and changed there, by way of persons. But how? With just the concepts of person, symbol, and situation of interpersonal interaction there is no way to tell.

To join the idea of a situation of interaction we need the idea of a system of interaction. Persons, after all, extend beyond situations into social systems to which they connect. *By "social system" we refer to a differentiable sequence of resource flows so functionally organized for persons as to constitute a dynamic process (or more than one) of evolving structures*—allowing that the process may be interrupted in time, dispersed in space, and intermittently applicable to given persons. Like persons, a social system is transituational.

A given social situation, by contrast, is multisystemic. It contains, because of its different acting persons, simultaneous episodes of interaction that are realizations of states of multiple systems. An interaction situation thereby serves as a locus for change in the various system states that the situation temporarily brings into confluence. The vehicle of change is the person, who connects the situation including its resource flows to one or more systems in which he or she is implicated. The person achieves this by the processing of relevant situational inputs. Persons bring systemic inputs to situations and export situational inputs to their systems. Along the way they transform the content of both, using their identity sets as anchors in terms of which they gauge the meaning content of the information that their conditions of interaction force them to process.

We emphasize that their conditions of interaction are twofold—situational and systemic—and the kind of interaction in which they engage is likewise twofold—symbolic and resourceful. How does all this sort out? It will raise no eyebrows to suggest that situational interaction provides a context for symbolic interaction. But having interpreted symbolic interaction as indicating potential resource flows for persons, to what can we assign persons' experience with active resource flows that are not symbolized at all? If the experience of persons is two-fold and transituational, their identity sets need to be organized so that persons can gauge their relations in networks of ongoing resource interactions, so as to respond to possible changes in their positions in those networks. Situations cannot provide them with that. Situations are *where* persons experience interaction, and the content of situations is inherently symbolic. Situations may provide a structured process useful for persons to interactively define the order of things, but the order there defined is not an end in itself. Persons can not survive just on symbols.

This leads us to suggest the following about the content of social systems: From the person's vantage point, a social system provides an organization of the resource interactions in which the person is implicated with others. That is to say, adopting a social psychological attitude, social systems should not be conceived in terms of persons nor symbolic relations between them nor structures of persons and their symbolic relations. *Rather, social systems should be conceived as networks of active resource flows between persons and structures of those networks, which the activity of persons generates.*[13]

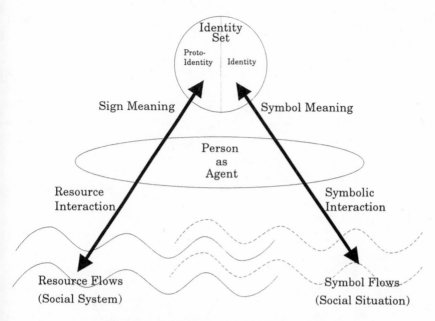

Figure 2. Identity set and social structure

There is no reason to conceive a system as a linear sum of interpersonal situations or structures thereof, and there is reason here to conceive it otherwise. Symbol meanings and sign meanings present for persons two different sorts of experience, the reflective and the immediate. But the "immediate" experiences of a person are not disconnected. If they were, the person would be disconnected. Persons extend beyond situations by way of their identity sets, which provide them with stores of information. *By using the information stored in their identity sets, persons may relate their situations of symbolic interaction to their systems of resource interactions.* This is portrayed in Figure 2.

It should be remarked of Figure 2 that we explicitly presume that changes in system resource interactions, insofar as persons are able to affect those, are initiated by means of the individual comparison process described by the Burke model. Each person compares resource flows in the situation with the expected flows maintained by their proto-identities, and in the event of some discrepancy takes action on the basis of that comparison. Similarly, symbolic interactions in the situation are the result of individual comparison processes, whereby each person compares symbol flows in the situation with expected flows maintained by their identities, and in the event of some discrepancy takes action on the basis of that comparison. The two flows—resources and symbols—are thus coordinated by being constantly and simultaneously processed by individuals. Each person simultaneously acts to adjust

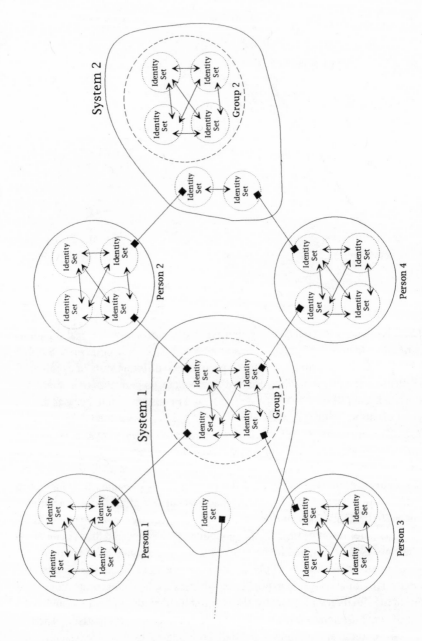

Figure 3. Systems and situations of interaction

flows of symbols and resources to match the flows that are expected in their identity sets. Identity sets are the mechanism which coordinate social systems of resource flows and situational interaction involving symbol flows.

But, of course, this sounds simpler than it is. A more accurate but complicated portrayal of the simultaneous interaction processes by which persons relate situations of symbolic interaction to systems of resource interaction can be found in Figure 3.

Figure 3 portrays two separate systems. While the resource flows defining such systems are not represented, the identity sets that control the resource flows through resource interactions are represented. System 1 is controlled by five identity sets while System 2 is controlled by six identity sets. Within System 1 is portrayed a group or situation of interaction consisting of four interacting individuals—specifically the system relevant identity sets of four individuals who are engaged in symbolic interaction. The four persons whose relevant identity sets are part of Group 1 are depicted around the system in the figure. Each of the four persons is shown with multiple identity sets. Four identity sets are shown for each, but that common number is arbitrary. What we intend to portray is that each person has a system relevant identity set and other identity sets not connected to System 1. The identity sets connected by box-end lines that extend from System 1 to each person are the same identity sets. Note that one identity set relevant to System 1 is not part of the situation of interaction of Group 1 because the person is not present to engage in symbolic interaction.

In System 2 there are six relevant identity sets. Four of these are currently engaged in a situation of symbolic interaction as Group 2. Two of the relevant identity sets are not part of the situation of interaction (Group 2) because the persons (Person 2 and Person 4) are currently physically present in Group 1. They may, nevertheless, engage in symbolic and or resource interaction relevant to System 2 while they are part of Group 1, thus illustrating both the multisystemic character of situations and the transituational nature of systems. On this latter point, one may think, for example, of two members from the same department attending a conference together. It is possible for them to conduct department business while in the conference.

CONCLUSIONS

All social action, structure, and process begins with personal activity. This personal activity is a comparison process for which the Burke (1991) model of identity provides parameters when suitably generalized. The generalization extends from considering personal identities in situations to personal identity sets, which are linchpins for persons to relate their situations to their systems. Persons must interpret potential resource flows in situational interaction to the active resource flows of their extended systems. They must interpret from systems to situations as well. They are presented always with a problem of correspondence: How does the array of potential resource flows, defined by the symbolic configurations of a situation, fit the systems of active flows in which the person is an agent? Will those active flows be changed—enhanced, depleted, suspended, reconstituted—and how? The person's position as a continuing agent in systems of active flows depends upon the correspondence between those active flows and the potential flows that symbolic interaction defines as relevant. The person's situational activity depends upon that correspondence also. A potential resource flow in a situation is a reference point for active resource flows in a system, and conversely. Resolving the problem of correspondence between the two provides the person with an orientation in situational interaction. The orientation is to sample the potential resource flows that the situation indicates can be expected, compare this with the standards in the identity set, utilize knowledge of that to resolve discrepancies for transituational systems of interaction, and act accordingly.

The impetus to change the systems of resource interactions in which persons are implicated resides in the situations of symbolic interactions that persons experience. Conversely, the impetus to change their situations resides in persons' systemic connections. All this has to be routed through identity sets. There is no impetus for any change if situational input does not disturb these meaning systems. It is the absence of perfect correspondence between situations that define potential resource interactions and systems of active resource functions, that disturbs identity sets and provides the impetus for persons to try to redefine their episodic situations so as to bring them into greater correspondence with their continually evolving systems. Their survival as persons depends upon that, because they are sustained by their systems not their situations.

Persons know this, if they do not cognize it, which is why they are proactive. Their structures of social interaction are their ways of addressing the problem of establishing a satisfactory correspondence between symbolic and resource interactions. To this end persons may create symbols or resources as needed, or destroy them as needed, where the "need" is gauged according to their identity sets. Their identity sets summarize their experience. This line of thinking suggests to us the following course correction for theorizing about interpersonal social interaction: For the evolution of microscopic structures it is not relations *between* persons that matter. Persons *are* the relations between the things that matter—the potential resources symbolized for them and the active resources signified to them. Persons relate the two by processing information through identity sets, which enable them to navigate through a sea of change, becoming agents who transform the symbol and resource flows they experience en route. Social interaction is organized according to that.

ACKNOWLEDGMENTS

We wish to thank Vik Gecas and Jan Stets for their helpful comments on an earlier version of this paper. Direct all correspondence to Peter Burke, Department of Sociology 4020, Washington State University, Pullman, WA 99164.

NOTES

1. We are using the term language here and throughout in a broad manner, consistent with studies in sociolinguistics, to include not only verbal language but also bodily and affective gestures, appearance and the like, to the extent that these are symbolized.
2. We are here working with the notion of identity as the set of meanings applied to the self in a social role that define what it means to be who one is.
3. This is an application of and possible derivative from Mead's perspective theory of objects (Mead 1938, pp. 159–165). In this theory, according to the principle of the perspective, an object exists only in relation to other aspects of its world or environment, which helps to define it. Because objects are socially defined and are relational, and identities are supposed to be objects, identities are supposed also to be relational.
4. Some semeioticians, such as Thomas Sebeok, consider it "impermissible" to ensnare the idea of a symbol with the concept of natural language (1975, p. 90). But symbolic interactionists have conventionally done this to a considerable degree. If the rhetoric in this paper seems to conform to the convention, it is not because we endorse the convention but, rather, because we think the limitations of conventional theory will be more apparent if we use it.
5. This is substantial understatement. Apart from ethological studies too numerous to cite, there now exists a field of inquiry, complete with its own professional organization, devoted to the study of biosemiotics.
6. Mead did not endorse it dogmatically either, except to insist that meaning is conscious only when it becomes identified with significant gestures, requiring symbols. Mead emphasized symbol meanings because only from these could there arise mind and self, which he cared to explain. Though he thought that significant gestures were available only to humans, he did not discount either the social origins or implications of unconscious meanings—as modern symbolic interactionists are wont to do.
7. This issue is very complicated, and the nomenclature in current use in social psychology—including ours—is not adequate to address it. For some complications see Walker (1983) and Bickerton (1990).
8. This term was suggested to us by Viktor Gecas.
9. We are not suggesting a sharp empirical distinction between signs and symbols. Signs can become symbols as their meaning becomes shared and as people are able to reflectively respond to them. Whether a particular stimulus is acting as a sign or a symbol is thus a difficult question. Theoretically, the distinction in clearer. As the stimulus functions to impart information about the current state of resources flows, it is a sign. See the discussion forthcoming.
10. To anticipate some of what follows: Our notion of resources is one of the resources in use. This is a dynamic concept which is best captured by the idea of resource transfers. Resource transfer refers to the movement of a resource from one place and time to another as it is either used or positioned for use to sustain individuals or interaction among individuals. The same idea lies behind the concept of resource flow. This movement and positioning of resources as it is structured and patterned is what, later in our analysis, we shall mean by a social system.
11. No distinction between resources and opportunities is required on this view. In distinguishing them, Emerson (1987, p. 21) uses the example of the availability of a bucket as providing an

opportunity to carry water, a resource. For his purposes, this works. But we want to know, for our purposes, why that bucket itself is not a resource, whey the bucket maker is not a resource. Is all this not resourceful?

12. Strictly, identity standards.
13. This argument was made by Freese (1988b), but without grounding it in a theory of the person. Where Freese took persons as givens, we take them as problematic.

REFERENCES

Bickerton, D. 1990. *Language and Species*. Chicago: University of Chicago Press.
Blumer, H. 1969. *Symbolic Interactionism: Perspective and Method*. Englewood Cliffs, NJ: Prentice-Hall.
Box, H. O. 1984. *Primate Behaviour and Social Ecology*. London: Chapman and Hall.
Burke, P. J. 1991. "Identity Processes and Social Stress." *American Sociological Review* 56:836–849.
de Waal, F. 1989. *Peacemaking Among Primates*. Cambridge, MA: Harvard University Press.
Emerson, R. M. 1987. "Toward A Theory of Value in Social Exchange." Pp. 11–45 in *Social Exchange Theory*, edited by K. S. Cook. Newbury Park, CA: Sage.
Fisch, M. H. 1978. "Peirce's General Theory of Signs." Pp. 31–70 in *Sight, Sound, and Sense*, edited by T. A. Sebeok. Bloomington, IN: Indiana University Press.
Foote, N. N. 1951. "Identification as the Basis for A Theory of Motivation." *American Sociological Review* 16:14–21.
Freese, L. 1988a. "The Very Thought of Resources." Paper presented at the Annual Meetings of the American Sociological Association, Atlanta, GA.
———. 1988b. "Evolution and Sociogenesis, Part 2: Social Continuities." Pp. 91–118 in *Advances In Group Processes*, Volume 5, edited by E. Lawler and B. Markovsky. Greenwich, CT: JAI Press.
Goodall, J. 1986. *The Chimpanzees of Gombe: Patterns of Behavior*. Cambridge, MA: Harvard University Press.
Harris, M. 1977. *Cannibals and Kings: The Origins of Cultures*. New York: Random.
———. 1979. *Cultural Materialism: The Struggle for A Science of Culture*. New York: Random House.
———. 1989. *Our Kind*. New York: Harper and Row.
Hartshorne, C. and P. Weiss. 1958. *Collected Papers of Charles Sanders Peirce*. Cambridge, MA: Harvard University Press.
Heltne, P. G. and L. A. Marquardt. 1989. *Understanding Chimpanzees*. Cambridge, MA: Harvard University Press.
Kinzey, W. G. 1987. *The Evolution of Human Behavior: Primate Models*. Albany, NY: State University of New York Press.
Lindesmith, A. R., and A. L. Strauss. 1956. *Social Psychology*. New York: Holt, Rinehart and Winston.
Mead, G. H. 1934. *Mind, Self, and Society*. Chicago: University of Chicago Press.
———. 1938. *The Philosophy of the Act*. Chicago: University of Chicago Press.
Peirce, C. S. 1958. *Selected Writings: Values In A Universe of Chance*. New York: Dover.
Sebeok, T. A. 1975. "Zoosemiotics: At The Intersection of Nature and Culture." Pp. 85–95 in *The Telltale Sign: A Survey of Semiotics*, edited by T. A. Sebeok. Lisse, Netherlands: Peter DeRiddr Press.
———. 1991. *A Sign Is Just A Sign*. Bloomington, IN: Indiana University Press.
Stryker, S. 1980. *Symbolic Interactionism: A Social Structural Version*. Menlo Park, CA: Benjamin Cummings.
———. 1987. "The Vitalization of Symbolic Interactionism." *Social Psychology Quarterly* 50:83–94.
Walker, S. 1983. *Animal Thought*. London: Routledge and Kegan Paul.
Weigert, A. J. 1983. "Identity: Its Emergence Within Sociological Psychology." *Symbolic Interaction* 6:183–206.

DISCRETE EVENT SIMULATION AND THEORETICAL MODELS IN SOCIOLOGY

Thomas J. Fararo and Norman P. Hummon

ABSTRACT

Simulation is a tool for the development of theories. Discrete event simulation is important for building dynamic models in which categorical state description and irregular changes of such states are key features. Processes in social networks and the micro-macro linkage problem are two general substantive concerns we consider using discrete event simulation (DES) methodology. Problems in social network analysis that relate to process divide into two analytical types. For one type, the structure of the network is given and the problem is to show how some process in the network generates various network-level outcomes. In this context, we show how DES methodology can be employed to help solve a problem in exchange network theory: namely, specifying the processes of interaction that lead from a given network shape to a distribution of resources. The simulation model embodies a dynamic account that explains comparative statics outcomes. We use this context to illustrate the implementation of DES methodology in a special English-like simulation language. A

Advances in Group Processes, Volume 11, pages 25–66.

second analytical type of network problem involves taking the network structure as problematic, as that which is to be explained by some processes of interaction. In this context, we show how DES methodology can be employed in a theoretical research program dealing with the problem of the formation of dominance structures in animal groups. Discrete event simulation implements the directive to "think process" in sociological explanation, with a focus on the concrete actions of actors in situations, while also implementing the directive to "think structure" in regard to the embeddedness of action in social networks.

INTRODUCTION

In recent years, simulation has made a comeback in sociology. Undoubtedly, the spread of personal computing and the acceleration of improvements in both hardware and software have been important and probably necessary conditions. The real impetus for the revival is the use of simulation by sociologists committed to theoretical model building. Collins (1988, Appendix), writing in the context of a comprehensive text on theoretical sociology, provides an argument for simulation as a significant methodology for theory development in sociology. Hanneman (1988) makes the more specific case for a particular form of simulation and an accompanying simulation language, giving extensive examples, a number of which are closely tied to classic theoretical ideas (e.g., those of Pareto and Marx). Heise (1986) has shown how simulation can be built into a long-term theoretical research program in which the fundamental dynamic assumptions, in his case, are embodied in a simulation of social interactions. Carley (1991) traces out the long-term consequences, in terms of group survival, of concatenated mechanisms that transform the cultural and social states of a group. Hummon (1990a) edited a special journal issue that presents simulation in relation to theory development.

Theories of more particular phenomena also have been areas of intensive development of theoretical models employing simulation techniques in an important way, such as research on crowd behavior (Johnson and Feinberg 1977, 1989; Feinberg and Johnson 1990; McPhail 1991) and research on power in exchange networks (Cook, Emerson, Gillmore, and Yamagishi 1983; Marsden 1983; Markovsky 1987; Yamagishi, Gillmore, and Cook 1988; Skvoretz and Willer 1991) and related social interaction processes, sometimes treated in the context of their value for teaching purposes (e.g., Bainbridge 1986; Leik, Anderson, and Gifford 1987).

Despite this growing significance of simulation in sociology, the total volume of such work and the necessary conceptual understanding and technical skills are still far below our needs for the immediate future. We need contributions in which the logic of simulation as such is highlighted and discussed in such a way as to contribute to its diffusion in the professional community of sociologists who are not afraid to cross the traditional boundary between theory and methodology. Put another way, we need what some sociologists (for instance, Freese 1980 and Fararo

1989) have been calling "theoretical methods." Very importantly, the wider community of sociological theorists should begin to get a sense of the possibilities for using simulation to develop their ideas.

Perhaps Hanneman's (1988) book is the exemplar contribution using simulation as a theoretical method. This paper aims to make a contribution of this sort. In fact, the aim of this paper is very close to that of Hanneman: to provide an exposition of one key type of simulation and its significance for theoretical sociology. Another common aim is to make the discussion concrete and operational through the explication and use of a particular simulation language. Yet there is a very significant conceptual distinction between the approach taken here and that of Hanneman. In this paper, a distinction will be made between continuous simulation and discrete event simulation. In these terms, Hanneman provides a lucid formulation of the logic of continuous simulation and of the simulation language DYNAMO that implements it, and he does so with examples that are recognizably sociological and theoretical. Analogously, we hope this paper will provide a clear formulation of the logic of discrete event simulation, and of the simulation language Simscript that implements it, and also offer examples that are recognizably sociological and theoretical.[1] Thus, the aim is to present the logic of discrete event simulation, as connected to problems of method in theoretical sociology and as implemented in a particular simulation language that illustrates the general principles of the approach.

To provide a substantive focus to the discussion, the methodology and its implementation will be linked to two types of theoretical problems connected with social networks. First, the network structure is a given and the aim is to explain some phenomena in its terms. The particular problem focus will be explaining outcomes of exchange processes in terms of the structure of constraints implied in the network shape. Second, the network structure is problematic, and we aim to explain how the structure arises. The particular problem focus will be on the explanation of how structures of dominance arise in animal interactions. Neither the particular problem foci nor the two categories of problems exhaust the potential role of discrete event simulation methodology in social theory, but they provide some definite instances of linkage between this methodology and problems of theoretical sociology.

In the first section, we conceptually describe discrete event simulation, placing it in the context of formal modes of representation of processes. In the second section, we introduce the methodology of discrete event simulation. We do this in the context of our example of the first type of network problem, wherein the network is given and we are accounting for some social phenomenon that varies with the network structure. Next, we show how this logic is embedded in Simscript. This section shows how a discrete-event theoretical model that generates offers and exchanges in a network is translated into a Simscript program. It will be evident that this simulation language has an English-like character that does *not* intrinsically involve postulating mathematical relationships among variables. Rather, and this is the core of its implementation of a discrete event simulation methodology, it involves directly representing activities and interactions, social events created by actors in social situations. The method will thereby illustrate Collins' (1988) thesis that simulation has a mediating role to play between the qualitative and the quantitative aspects of our field. Further, this point will connect with the micro-macro link question explored in recent sociological theory (Alexander, Giesen, Munch, and Smelser 1987; Coleman 1990). At the level of *derived outcomes* there is a conceptual location for relating systemic properties to each other, while at the level of generative process we have the interdependent actions of acting units.

In the fourth section, we show how discrete event methodology and the use of the Simscript implementation of it can function in a theoretical research program. As indicated previously, the focus is now on taking network structure as problematic, as a phenomenon to be generated over time through social interactions. We discuss two families of models in terms of their specific theoretical formulations, the nature of the process rules they postulate, and the way in which discrete event simulation helps develop the ideas in the program. Finally, the last section reviews some of the major points in the paper and looks toward further developments in this area that are currently being pursued in ongoing research.

DYNAMIC MODELS AND DISCRETE EVENT SIMULATION

Simulation is above all concerned with process. We use the definition that "*a simulation is a dynamic model implemented on a computer*" (Evans 1988, p. 19, emphasis added). If the properties of a process, as represented in a model, were obvious or could be derived by analytical (i.e., mathematical) methods, then there would be no need for simulation. The need arises because scientific experience reveals again and again the following aspect of dynamic models: simple rules have complex consequences. We cannot readily tell, in many cases, what the rules imply by just thinking about them. The rules, or mechanisms, can concatenate in complex ways over time, so that the outcomes can be quite varied, depending upon initial and parametric conditions or inputs. Thus, to "implement" a dynamic model on a computer is to create a computational model that mirrors the analytical model such that computational processes will produce the outcomes of the model. As we vary conditions, model outcomes vary and these arise by virtue of the time-dependent concatenation of the processes formulated in the simple rules that define the model. To repeat: a simulation is a dynamic model implemented on a computer in order to study the properties of the model that cannot be intuited or derived directly from its postulational basis.

In order to appreciate the dynamic nature of simulation, consider the set of choices we face when we construct formal theoretical models. First, from a substantive viewpoint, explanation is the aim of a theoretical model and explanation is best attained through specification of one or more *mechanisms* or processes that produce the phenomenon to be accounted for. Thus, the explanatory function of theoretical models takes us directly to dynamic models. Second, however, we have options in creating such models. We can put these options in the form of a set of menus, each menu containing a list of possible variants. We first list the whole set of menus and then discuss each of these as a prelude to our simulation focus. The key menus are:

1. State space: categorical, continuous.
2. Parameter space: categorical, continuous.
3. Time domain: discrete, continuous.
4. Timing of events: regular, incessant, irregular.
5. Generator: deterministic, stochastic.
6. Postulational basis: equations, transition rules.

A formal-theoretical model of a process requires a representation of all of these elements, involving some pattern of choices from the menus. Our next step is to describe the choices from each menu and to characterize discrete event simulation in terms of the overall pattern of choices.

State Space and Parameter Space

These choices concern some set of units or entities, in various relationships. The entities are characterized in terms of properties that may vary in time. Such a time-variable property is an element of the state of the system. The state space is the complete set of combinations of all possible values of all the dynamic variables. The two choices in menu 1, then, are essentially related to measurement level. In the categorical case, we are treating qualitative features of the world, whereas in the continuous case, we are treating quantitative features.

A fixed property of a unit or of the system of units is a parameter. Menu 2 applies the same conceptual distinction as menu 1 but now applied to the conditions, the configuration of parameter values, under which some processes of change in state variables occur. The two menus, taken together, define the formal setting for the study of dynamic phenomena (for further details, see Fararo 1989, Chapter 2).

Representation of Time

Menus 3 and 4 concern the representation of time. The usual formal way of distinguishing models is via the nature of the time domain and menu 3 formulates the two basic choices. If the state variables are defined and analyzed only in terms of discrete time points labeled by integers (e.g., year 1, year 2, . . .), we have a discrete time model. If the state variables are defined for every moment of time, then we have a continuous time model. For instance, the famous Asch (1951) study consisted of a series of rounds and in each round the behavior of the naive subject was recorded. It was natural to model the situation as one involving a discrete time domain, with time t identified with trial t (Cohen 1963). By contrast, as Coleman (1964) pointed out early in the history of mathematical sociology, since sociology largely treats non-contrived processes that are continuously occurring in the world, theoretical models of processes are most naturally of the type in which the time domain is continuous.

A second feature of the representation of time is what we call "timing of events." For present purposes, we define an event as a change of state. There are three possibilities. First, events can be *regularly* scheduled, as in many institutionalized action settings when certain rituals are performed at certain definite intervals (e.g., graduation ceremonies in academic settings, involving change in actor status, the latter viewed as a categorical state). Second, events can occur *incessantly*, as in the continuous change of position of a moving object in physical space or (in an idealized model) the changing state of populations in an ecosystem. Third, events can occur at *irregular* intervals of time. Numerous social actions and events occur in such a way: people arriving and departing from waiting lines, encounters between people at work or in a neighborhood, conflict episodes between collectivities, and so forth.

Generator and Postulational Basis

By the generator of a process, we mean the formal representation of the mechanism, or combination of mechanisms, that produces changes of state "in the small." Long-range transformations are produced by iteration of the generator, reapplying to the very state it generated. In discrete time, application of the generator to the current state, at time t, transforms that state into the next state, at time $t+1$. If $X(0)$ is the state at time $t = 0$ and if the generator is denoted by the transformation T, then:

$$X(1) = T(X(0))$$

$$X(2) = T(X(1)) = T(T(X(0))$$

$$X(3) = T(X(2)) = T(T(T(X(0)))$$

Thus, the "long-range" change of the state of the system from time $t = 0$ to time $t = 3$ is given by the three successive applications of the generator T, as shown above in the last expression on the right. In continuous time, the same logic of in-the-small and long-range transformations holds. For instance, differential equations are specifications of the "in the small" generator and the integration of the equations, analytically or numerically, is the "long-range" element, often called the solution of the equations.

When the generator is deterministic, transitions occur with probability one between pairs of states in the state space, either in discrete or in continuous time. When the generator is stochastic, such transitions occur in a probabilistic fashion.

Hence a deterministic generator is a special case of a stochastic generator, conceptually.

Menu 6 refers to what we call the postulational basis of a theoretical model. Probably most sociologists think of models as specified by equations. Indeed, this is one important and very significant mode of postulation of theoretical models. Classically, theoretical process models have been framed as systems of differential equations (continuous time) or systems of difference equations (discrete time). But an alternative exists. The theoretical model might be framed as a set of rules that enable formal analysis but in which there is no postulation of equations. For the categorical types of state space, clearly, this sort of postulational basis is critical and the development of formal theoretical analysis in sociology must include work that enables dynamic analysis involving qualitative aspects of social phenomena. It is this feature that is central to our approach and to this paper.

To make this point clear, and to relate it to alternatives in the simulation of social processes, consider how differential equation models arise. The basic requirement is that the time domain be continuous. This is only necessary, not sufficient. The state space and generator aspects have to be considered. Roughly, there are two contexts in which differential equations arise in continuous-time models: continuous state variable deterministic contexts and discrete state variable stochastic contexts.

In the first context, the "motion" of the system is a smooth curve in state space corresponding to incessant events of change of location of one or more units. In classical mechanics, the generators of such smooth curves all are instances of a basic differential equation template associated with Newton. Recall that a simulation is a dynamic model implemented on a computer in order to study the properties of the model, that is, to discover what it implies about the processes modeled. In the case of differential equation models in this classical context, the type of simulation is that advocated by Hanneman (1988). It is usually called *continuous simulation*. The continuous time line is broken up into very small intervals, and successive computations over these intervals produces the long-range integration from the equations specified in the small. Simulation is essentially a method of calculation of numerical solutions of a system of given equations. Timing involves incessant events, that is, incessant changes of state.

The second context is illustrated by Coleman's (1964) focus: a set actors such that actor categorical states (such as orientations toward a candidate, for or against) exist at each moment of time. Change of state can occur by virtue of influences from other actors, embedding the process in a social network. The state of the system at any moment is the number of actors in each categorical state. In this type of model, the initial menu choice involves stochastic transition rules between categorical states (for a statement of the rules, see Fararo 1973, Section 13.2). The model thereby defined is sufficiently simple to enable the derivation of a system of differential equations for the change in the probability distribution over the possible system states. Given such equations, they can be used to generate transitions by the use of a random number generator at each appropriate point in the process. With a large number of repetitions, the statistics of the process are obtained. This is called Monte Carlo simulation. It involves no conceptual novelties or sociological content. Like numerical integration in the first context of differential equations, it is simply a tool for computing results.

However, in this second context of transition rules, another type of situation can arise: rules can be posited but equations cannot be derived. For theoretical sociology, this is perhaps the most common type of analytical situation. We have some picture of a complex of interdependent and dynamically varying situated actions taken by diverse actors that, under varying conditions, lead to distinct outcomes. We want to represent very explicitly the relevant social events as these occur and create new conditions for the actors over time. Yet, without equations to be solved, how can we derive the outcomes? This is the question to which discrete event simulation is a proposed response.

We are now in a position to define the special conceptual character of discrete event simulation. First we recall various points already made: we are concerned with theoretical models; these have the function of explanation; explanation requires specification of mechanisms or processes; this leads to an interest in modes of construction of dynamic models; a simulation is an implementation of a dynamic model on a computer so as to derive its analytic properties; we use menus of modeling choices to facilitate our understanding of constructing dynamic models. Now we present our definition of discrete event simulation.

A discrete event simulation model is a dynamic model implemented on a computer in which there is a system of interdependent entities and their activities described in terms of the following: *categorical states*, a *continuous time domain*, with events occurring at *irregular times*, and a postulational basis of *transition rules* among the categorical states.[2]

Discrete event processes are different from classical continuous time processes. They occur in continuous time but have a temporal discreteness property that distinguishes them from classical continuous simulation models that correspond to deterministic systems of differential equations. The key difference lies in the timing of events: although events *may* occur at any time, they do not occur incessantly but at separated, irregularly spaced times. The events in question involve the actors and their acts, decisions or activities. An act is a process bounded by two discrete events: its start and its end. Because the *change* of state is not incessant, the models are in continuous time but the acting units only change state at discrete time points. This is what gives rise to the terminology "discrete event."

Discrete event processes are also different from discrete time processes. Discrete event processes produce a record of times of events as an aspect of each of the events generated. For example, consider a hospital emergency room. People arrive randomly after some tacit or explicit triage phase, and enter a waiting phase in an emergency room, waiting to see a physician. We can represent the people in this room as a queue. The queue has attributes of the number of people, and an order.

Table 1. Emergency Room Event History

Time	Event	Number of Waiting Patients
t_1	Patient arrives	5
t_2	Patient arrives	6
t_3	Physician sees next patient	5
t_4	Physician sees next patient	4
t_5	Physician sees next patient	3
t_6	Patient arrives	4
t_7	Patient arrives	5
t_8	Physician sees next patient	4
t_9	Physician sees next patient	3

The second entity of this system is a physician who examines patients for varying lengths of time. The dynamics of this system, starting with four people waiting at time t_0, might evolve as described in Table 1.

Each event in Table 1 causes a change of state of the system, for example, the queue changes. These are discrete events: they occur in continuous time, but irregularly at discrete instants where the state of the queue changes.[3] In this example an important aspect of the process, from both the actor and the system viewpoints, is the average waiting time before seeing a physician. This time depends upon the interarrival time distribution and the physician "service" time distribution. If we modeled the emergency room by using ordinal times of arrival of people to create discrete times 1, 2, 3, . . . , corresponding to the first arrival, the second arrival, the third arrival, . . . the model could represent the changing state of the queue to see a physician but it could not enable us to compute such critical quantities as the average waiting time. This computing loss would arise, conceptually, because we had tried to model a continuous-time process with irregular events as if it were a simple discrete time model.

METHODOLOGY OF DISCRETE EVENT SIMULATION

Having defined what we mean by a discrete event simulation model, we now proceed to a generalized characterization of the methodology of constructing such models. Hereafter, we write DES for "discrete event simulation." To help the reader follow the presentation, we will illustrate the concepts in the domain of social networks. In particular, as previously described, there are two analytical contexts in theoretical models. In one context, the network is given and we are explaining some social phenomenon that depends upon its structure, whereas in the other case, the network structure itself is to be explained. In this section, we employ an example of the first analytical context; in a later section, we employ an example of the second analytical context. We begin by providing some background for the definition of our theoretical problem.

Exchange Networks: The Process Problem

The particular research program to which our work relates involves power in exchange networks (Willer 1992). In such theories, the social network is construed as a system of potential interactions in which exchanges occur involving resources. In a class of cases of interest, each actor occupies a position in the network and is constrained to exchange with a subset of the possible exchange partners, each possible partner defined by a preexisting and fixed link in the network. Within this class of cases, an important initial and relatively simple case, chosen to initiate both experimental and theoretical analyses, involves the choice of a single exchange partner, thus excluding all others from the particular exchange. Since the distribution of resources that emerges out of such exchanges depends upon the patterning of alternatives of exchange for each actor, the problem fits the paradigm of explaining a phenomenon (the exchange outcomes in terms of distribution of resources) in terms of a given social network.

Our example is constructed in relation to experimental configurations studied empirically by Skvoretz and Willer (1991). In this work, the problem is to predict a resource distribution that arises through a process of offers and counteroffers made by people occupying positions in an experimentally established small communication network. By varying the channels that are open or closed between each pair of nodes in the network, different structural shapes can be empirically realized as parameters of exchange processes. People functioning as subjects in such an experiment sit at computer terminals and receive controlled information on the display, which shows each of them their current list of offers, both sent and received. Additional information also may be displayed. The task of each person is to arrive at an exchange by making and responding to offers that take the form of a proposed split of a certain experimentally defined resource, usually called "points." For instance, person A may propose a 13–11 split of 24 points to person B, who may have a 14–10 offer from person C currently under consideration, and so forth. When an exchange between a pair is agreed upon, that pair is deactivated from the network. The session ends when no further exchanges are possible. Not everyone necessarily enters into some exchange. For instance, with five nodes in a line, one node gets left out. The situation is repeated a number of times so that at each position in the network, the average amount per exchange earned at that position can be calculated. Positions that are favored, in some sense, by the shape of the network should earn more points, on average. Individual ability in such matters accounts for variations but the structural effect of the network position is reflected in the averages with different individuals occupying the position. Thus, the experimental analysis takes network shape or structure as independent variable and the resources earned at each position as the dependent variable. The goal of the analyst is to show how the latter is predictable from the former. An example of a network shape is shown later in Table 2, which is used in our example of a DES model later: it is the "5-Node Line Network," with the interpretation that each actor can make and receive offers from an actor occupying an adjacent network position.

One mode of analysis leading to experimental predictions involves a "graph-theoretic power index" (GPI) calculated for each position in the network (Markovksy, Willer, and Patton 1988; Willer, Markovksy, and Patton 1989), a measure that takes into account indirect ties and so the alternatives that an actor's prospective partner has to exchanging with the actor. For example, for the 5-node line network, the GPI of nodes 2 and 4 is 2, while that of nodes 1, 3, and 5 is 0. So, the "even" nodes should gain more in exchanges with their neighbors.

This GPI-type of analysis involves a hypothesis of the comparative statics type: variations in a parameter (GPI) produce variations in a social phenomenon (resource allocation). This is valuable but differs from our approach, which puts primary emphasis on the representation of the mechanism that generates the hypothesized relation between the structural givens and the outcomes of exchange. This is our point of departure: the theoretical problem to which our DES model is addressed is this *process problem* in exchange network theory.

Micro-Macro Linkage and Adaptive Rationality

How shall we approach this process problem? The fundamental feature of our model will be a process specification of a conceptualized micro-macro linkage. In terms of general conceptual meaning and theoretical postulate:

- Macro level: the state of a social system.
- Micro level: the action by an actor in a situation.
- Principle: action is adaptively rational.

Adaptive rationality is a process-oriented type of theoretical idea implementing a conception of rational action by human beings in situations. Envisioning a series of interactions rather than a static choice situation, the principle (Fararo 1989, p. 224) is that "the probability of an action changes with its consequences in terms of the sanction significance of those consequences, whether positive or negative."

In the present theoretical model the three general elements are applied to a specific, although still abstractly specified, context. Namely, the structure of the exchange network is the given macro condition. By scope assumption and experimental realization of this assumption, the structure is given, a fixed parameter. The micro level consists of acts in the situation defined by the task of arriving at some exchange with one of a number of actors to whom a given actor is connected in the network. Acts include offers, rejections, acceptances, and so forth. The application is relatively straightforward. When a 14–10 offer is rejected by alter, for the actor this reaction is a consequence with negative sanction significance. To attain the goal, but adapt to alter's response to this offer, the actor modifies the implied aspiration of 14. By how much? We assume the rational adaptation: from 14 to 13, rather than 12 or less, in order to earn "as much as possible." That is, the rational adaptation is downward by the minimal unit by which alter might accept another and lower offer. This is a deterministic adaptive rule, a special case of the general principle, since the best possible adaptive action is selected with probability one. By applying this rule over and over, we will generate the exchange network outcomes and show how they depend upon position in the network.

In terms of the metatheoretical micro-macro formulation suggested by Coleman (1990, Chapter 1), our DES model will start at a macro level (the given network), define the situation, including inducing initial levels of aspiration, for actors by their linkages in the network (macro to micro link), generate adaptively rational acts by these actors based on their current levels of aspirations (micro to micro link), and so generate exchanges that lead to a patterning of resource allocation by position in the network (micro to macro link). In this way comparative statics propositions relating a given network to patterning of outcomes of exchange in it are explained by the linkages incorporated into the model. Figure 1 presents a version of this micro-macro linkage in a somewhat truncated form (because the dynamic interactive aspect is not apparent in the figure).

Discrete Event Simulation and the Process Problem

Having described the background for the type of network problem of interest in this and the next section, we now proceed to our main task. We will present discrete event simulation methodology by reference to a DES model that addresses the process problem of exchange network theory.

The basic components of a DES model are *entities with attributes*. The concept of entity is sufficiently general to include not only acting units but also ties among such units. For instance, an attribute of an actor may be an activity status. An attribute of a tie between two such actors may be a bond-state, such as weak or strong.

In the exchange network DES, there are two types of entities: persons in network positions and the offers (including counteroffers) they make to each other. The first type is termed *actor* (or sometimes, *node*) in the DES model and the second type is termed *offer*. Actually, since offers can be made or received, there are two types of the latter in the model: *offer.made* and *offer.recd*.

In general, entities may be *permanent* or *temporary*. These terms are relative to the time domain of the model. To be temporary usually will mean "subject to birth and/or death," while to be permanent means to necessarily (as a scope feature) endure throughout the history generated by the model. In the exchange network DES, the actor-nodes are permanent entities, enduring from the start to the end of the experimental situation. The offers are temporary entities. When an actor in a

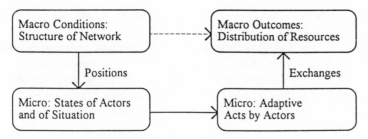

Figure 1. Micro-macro linkage

network position decides to make an offer, an offer is born or, as we shall say, is *created*.

Entities have attributes. In the exchange network DES, the attributes of actors include (but are not limited to) *level of aspiration* (as to the amount to be gained for the self in an exchange) and the *number of ties* to other nodes. The attributes of an offer include two actors, the initiator and the receiver, and the resource distribution it proposes between them.

Attributes of permanent entities have an aspect analogous to the permanent-temporary distinction among entities. Namely, permanent entity attributes are either time-invariant or dynamic. We fix the former for the duration of the DES and the latter change due to events in the DES. A constant may be called a *parameter* and a dynamic attribute may be called a *state variable* of the corresponding entity.

In the network exchange DES, the network structure is fixed and implies that the number of ties of an actor is a parameter. An example of an actor attribute that is a state variable is level of aspiration, since in the DES this aspiration can vary over time. The initial levels of aspiration are set when an assumption is made about the knowledge concerning the network that is available to the persons functioning as subjects. In the DES model being discussed, it is assumed that they know the number of others with whom they may exchange, but do not know the overall shape of the network. Thus, the initial level of aspiration is set by a nondecreasing function of the number of alternative exchange partners.

Recall that we define an event as a change of state at an instant of time. A state variable shifts value at a definite time point. We treat the level of aspiration as subject to such change in our DES exchange network model. An offer is generated in terms of some current level of aspiration. If the offer is rejected, an event of change of level of aspiration is generated. With adaptive rationality, the new level of aspiration will be the one that reduces by one resource unit the amount that the actor hopes to attain through exchange. This is the rule embedded in the model. We may take this as an example of a *transition rule*.

As this example makes evident, *the timing of events depends upon the activities of entities*: in this case, the action of one actor (the rejection of an offer) leads to a change of state of another (a one-unit reduction in level of aspiration). If the target were to accept the offer this event would not occur; if the target were to delay deciding and then later reject the offer, an analogous event occurs but at a different time point. That is, every event has a unique time-of-occurrence aspect as well as its associated change-of-state aspect and because these events are outcomes of actor choices that are neither scheduled for definite time points nor incessant—people pause to think and compare—the timing of events is of the irregular type.

Toward a Program Embodying The Model

In DES methodology, there are two steps through which the previously stated ideas become incorporated into a program embodying the theoretical model and so constituting a simulation model. The first step is *declaration* of the entities and their attributes. The second step is *writing the routines* that involve the dynamics of the processes. The routines involve the creation of definite model entities of the types set out in the declaration, the setting of their parameters to particular values, the initialization of the values of their state variables, and, of course, the representation of the processes through which the entities interact and create events that constitute the dynamic processes being simulated.

After we write and debug the program, we investigate the properties of the DES model. The typical procedure is to explore a variety of parametric conditions so as to see how the state variables behave under varying conditions. A particular and important special case involves studying how any implied equilibria of the DES model vary with the parametric conditions. In short, we vary the ways in which a dynamical system model can be studied to learn about the properties of a social process.[4]

These programming steps of DES are best appreciated in the context of a particular programming language. In the next section, we show how the DES methodology is implemented in one such language environment, continuing with the exchange network model as our example.

DISCRETE EVENT SIMULATION USING SIMSCRIPT

A simulation is a computer implementation of a dynamic model. Two basic alternatives exist in the choice of a computer language for simulation purposes. First, one can employ a general purpose language, such as BASIC. Second, various special languages have been developed to implement discrete event simulation methodology, as well as other types of simulation (Evans 1988). Each alternative has certain advantages and disadvantages. A general purpose language may have a wider base of users as well as more flexible technical characteristics. A special language may be especially adapted to the simulation task, reducing the amount of time required to program particular models. We are exploring both types of implementations of DES methodology.[5] In the present paper, our focus is on the second general option, the use of a special DES simulation language. Such special languages automate much that would otherwise consume valuable research time in terms of software design and implementation. And there are other reasons, to be discussed shortly. Simscript is one such special DES language. In this section, the aim is to show how the DES methodology outlined in the prior section is implemented in Simscript.

To use Simscript, one must acquire the software package by that name, including its manuals, and install the program on a suitable computer hardware configuration.[6] One advantage of Simscript is that it exists in a personal computer version. Another and important consideration is that the language is English-like so as to facilitate its use by scientists who are not professional computer programmers.

The Simscript language functions as an environment in which particular DES models can be developed through creating, testing, and using Simscript programs. A program is a set of *modules*. The modules are written in the Simscript language, "compiled" (translated into computer machine language) by the Simscript compiler and then integrated or linked into one overall program.

The modules are of various types that reflect the methodology of DES. One type of module, called the *Preamble*, is defined at the start. It contains the set of declarations of types of entities and their attributes, along with other Simscript-specific declarations. Every Simscript program must begin with a Preamble. A second type of module is set out immediately after the Preamble and is called the *Main* program module. It initializes parameters and state variables and starts the simulation by calling upon a built-in routine, called the *timing routine*, that automatically takes care of all the critical details involving time. Also, the Main module may include a loop for repeating a number of independent simulations, called repetitions or replications.

For instance, exchange network experiments with five subjects in five network positions may be repeated with 25 sets of five persons each. Simulation results are generally summarized for each such replication and for the set of all such replications in terms of some statistical summary. In other applications, the repetitions are not experimentally controlled but correspond to distinct observations, for instance, of distinct groups exemplifying the same generic system of processes the model is about.

Every Simscript program must have a Preamble and a Main module. Other program modules may be written and called from the Main program. In the language of programming, like Main, they are routines. Two special types of routines are critical to the logic of Simscript. These are written and called by the Main routine and perhaps by each other. One type is an event routine. The other type is a process routine. Generally, each type of event or type of process in the DES model will be translated into one such routine. Each time such an event or process is to occur, the corresponding routine is called. The important point is that the type of event or process is a *recurrent* event or process. An event is instantaneous, whereas a process will take up some interval of time. A process, in this sense, involves a recurrent *series* of events, so that an event is really a special case of a process with only one event. An empirical action is a process but it also might be represented as an event, depending upon the requirements of the model. In the latter case, the action is represented as simply occurring at some time point, without representation of the block of time it takes.

A type of recurrent event or process has several aspects in a Simscript model:

- It is declared in the Preamble and given a name.
- There is module with that same name which is the routine.
- There are an indeterminate number of process instances of that recurrent type that occur in the simulation.

To summarize so far in this introduction to the Simscript implementation of DES methodology: It is a specific language for DES model building. Every Simscript program is a set of modules that one writes using the language rules. A typical program consists of two necessary modules, the Preamble and Main, followed by one or more Process or Event modules, each corresponding to a distinct recurrent process or event.

Henceforth, we let NETWORK refer to the Simscript model that implements the exchange network model described in the prior section. Figure 2 sets out the basic activities represented as discrete action events and the transitions among them.

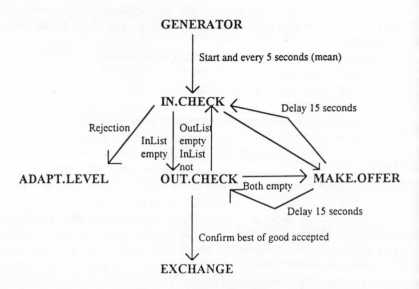

Figure 2. Network program

As a program, the structure of NETWORK is given in terms of the six modules that embody the six actions of Figure 2 as well as the Preamble and Main modules. In what follows, we introduce these modules, then present and explain sample program statements. Each module is bounded by a module name and a closing term END.

Modules

The Preamble Module

We declare the basic entities (actors, offers.made and offers.recd) and their attributes. One entity, the network, is not localized to any single entity and Simscript allows us to declare it as an attribute of the system. Part of the Preamble module is given in Listing 1.[7]

These statements illustrate three facilities built into Simscript. First, the Simscript concept of *set* represents a dynamic population of entities. Entities may own sets and belong to sets. The Preamble declares that each actor owns two sets: a set of offers made *to* that actor (IN.LIST) and a set of offers made *by* that actor (OUT.LIST). The temporary entities of the type offer.made belong to OUT.LISTs, whereas the temporary entities of the type offer.recd belong to IN.LISTs. In the experimental versions of exchange networks, such as the set-up devised by Skvoretz and Willer at South Carolina, these concepts refer to external memory devices: each actor sits at a computer monitor on which incoming and outgoing offers are displayed. For nonexperimental realizations, the concepts refer to an idealized version of internal memory (since no forgetting occurs in this model).

```
PREAMBLE
        Processes include GENERATOR
        Events
                Every IN.CHECK has a checking.actor
                Every OUT.CHECK has a reading.actor
                Every MAKE.OFFER has an offer.actor and a target.actor
                Every ADAPT.LEVEL has a rational.actor
                Every EXCHANGE has an offer, a first.actor and a second.actor
        Permanent entities
                Every ACTOR has an INIT.LEVEL.ASP, a NUM.TIES, a NUM.RECD
                        ...(and) owns an OUT.LIST, an IN.LIST, ...
        Temporary entities
                Every OFFER.MADE has an ID.NUM, a TO.NODE, an AMOUNTX, ...
                Every OFFER.RECD has an ID.NUM, a FROM.NODE, ...
        The system has a NETWORK
        Define NETWORK as an integer, 2-dimensional array
        :
        Tally AVG.NUM.RECD as the average of NUM.RECD
END
```

Listing 1.

The second Simscript facility the Preamble illustrates is Tally. This is a built-in routine that calculates statistical properties of state variables. At the end of every replication, each actor will have received and sent a certain number of offers, acting in a certain position in the network. The model will imply variability in a given position, because of the stochastic processes involved, so one wants to obtain averages. Tally does this, automatically averaging the appropriate attributes of the actors, over all replications, so that we can print it at the end as part of our simulation summary. In addition to averages, one can automatically obtain complete frequency distributions and other implied statistical properties.

Finally, the third Simscript facility shown in this Preamble is "the system" as an entity with certain attributes. Sometimes we have properties that are not localized attributes of acting units. The whole network is such an element. One possibility is simply to represent each actor's ties to others as an attribute of that actor. But Simscript provides an alternative mode of representation, used here. This is the idea of a single unique entity (called "the system") that can have certain attributes and own sets. In this model, a dynamic version of the active network (with exchanging pairs dropping out over time) is defined as an attribute of the system.

The Main Module

The actual execution of a Simscript program begins with the first line of this module. It functions to read inputs, initialize parameters and initial states, and to loop through a series of replications of the simulation by activating a GENER-ATOR. Later, when a simulation run or set of runs is done, it reports outcomes. Listing 2 reports portions of the Main module.

```
MAIN
    :
    Print 1 line thus
PLEASE ENTER THE NETWORK SIZE AND NUMBER OF REPLICATIONS
    Read N, NO.OF.REPLICATIONS
    :
    Create each actor(N)
    :
    For replication = 1 to NO.OF.REPLICATIONS
            Do
                    : (initializations)
                    Activate a GENERATOR now
                    Start simulation
                    : (Print results of one replication)
            Loop
    : (Print simulation summary for all replications)
    END
```

Listing 2.

Several elements of this Main program are of general significance. First, note the "create" statement. Recall that actors are the permanent entities of this model. This means that their number is fixed for the duration of the dynamic processes that unfold. Of course, this is a scope condition (Cohen 1989). Real exchange networks may have turnover. Experimental methods provide realizations of the networks with this scope condition satisfied. When "create every actor" is executed, Simscript reserves memory for the attributes required to characterize each of the N actors. By contrast, temporary entities are created one at a time, as they are generated in the dynamics that unfolds and the corresponding statement (e.g., "Create an offer") only reserves as much memory as one entity of the indicated type requires, given its attributes.

Second, consider the "activate . . ." and "start simulation" statements. These relate to the way that Simscript has a built-in simulation routine. There is a simulation clock that starts at zero. It is represented by a Simscript variable denoted time.v, so initially time.v = 0.0. The "activate" statement "schedules" an event or process instance of the given type by placing it in a queue. There is a time of activation and it can be "now," which simply means at the time given by the simulation clock variable, time.v. So the activate statement here places an instance of Generator in the queue with its time of activation as time zero. The statement "start simulation" activates a built-in routine that goes to the queue and finds the earliest event or process, executes the corresponding routine (with proper entities and their states) and then returns to the queue again until no events are found, when control passes to the next statement, usually leading to a printing of the results.

The GENERATOR Module

This is a routine that starts the offers going and includes a kind of refresher loop of spontaneous activations of IN.CHECK. Listing 3 shows portions of the Generator process.

Until all possible exchanges are done (which is a status assigned in EX-CHANGE), we execute a loop. An actor (called "new.actor" here) is generated at random from among those who are still active. An instance of IN.CHECK is

```
Process GENERATOR
    :
    Until end.test = done
        Do
            :
            Activate an IN.CHECK giving new.actor now
            Wait exponential.f(5,3) .seconds
        Loop
    END
```

Listing 3.

activated with this actor and with its time of activation the current simulation clock time, time.v.

Then the following statement illustrates another built-in Simscript facility: "Wait . . ." This means: the GENERATOR process instance, now being executed, is put back in the queue with its time to occur reset by sampling from an exponential distribution with mean of 5 seconds. For instance, the number 4.5 might be generated from this waiting time distribution. Then the next time for this GENER-ATOR to be recalled will be set as time.v + 4.5 (seconds).

When the timing routine goes to the queue and finds this instance of GENER-ATOR with this time of activation as the earliest event in the queue, *it advances the simulation clock to this time*. This is the time advance mechanism that is a key feature of DES models. The time at which the next event starts is used to move the simulation clock time to this point *without literally traversing each instant in between*. Provided that we are not yet done, when this instance of GENERATOR continues its execution, we activate another IN.CHECK "now": precisely the advanced time (since the clock time.v has been updated).

The IN.CHECK Module

This is a routine that represents an actor's scanning incoming offers, categorizing them as bad (less than current level of aspiration) or good (at or better than current level of aspiration). Various transitions are contingent on what is found, if anything, such as a transition to OUT.CHECK (no incoming offers) or a transition to MAKE.OFFER (replying to a particular bad offer that has been rejected). For each bad offer newly found by an actor, that offer is rejected. Listing 4 shows portions of the IN.CHECK event.

The first conditional transition of action is shown for no incoming offers to the particular actor, called checking.actor here. The second conditional transition of action is not given in detail, but what is happening is that the actor may find several offers from others in the network, all bad but some better than others. The actor

```
Event IN.CHECK given checking.actor
        :
        If IN.LIST(checking.actor) is empty
                activate an OUT.CHECK giving checking.actor now
                return
        endif
        :
        If ... (no good offers, but some bad offers rejected)
                let alter = from.node(best.bad)
                activate a MAKE.OFFER giving checking.actor and alter now
        endif
    END
```

Listing 4.

locates the author (called alter) of the best such bad offer (called best.bad) and activates a counter offer: we activate a MAKE.OFFER instance "now" for this actor to alter.

The OUT.CHECK Module

This is a routine that represents an actor's monitoring the status of offers in OUT.LIST (that actor): the offers.made of that actor. The actor looks for acceptances. The best of these forms the basis of an activation of an EXCHANGE event. This occurs as follows, according to the model embodied in the program: The actor constructs a temporary memory set (we call out.check.set) in which are placed all the offers.made that have been accepted by actors that are still active. The set is ranked by how good the offer is from the actor's point of view. If this set is empty, the actor does one thing, if not something else. In particular, below we show the code for what happens when this set is not empty, some acceptance exists. Listing 5 shows the OUT.CHECK event.

If the out.check.set of this actor (called reading.actor) is not empty, there is at least one acceptance. The actor finds the best of them: f.out.check.set, the first in the set from the actor's point of view. This offer is denoted best and then confirmed. Then an EXCHANGE occurs, giving the offer agreed upon and the actors.

The MAKE.OFFER Module

This routine creates offers. There are two versions, one for the OUT.LIST of the offer.actor and the other for the IN.LIST of the target.actor. The important point is that the amount proposed for the offer.actor is given by that actor's current aspiration state, with the other getting the remainder. The arbitrary total resource points of 24 is used because it corresponds to a value most often used in exchange network experiments. It is useful to display the offer as output in order to get dynamic output to check on the degree to which the simulation model produces plausible sequences of interaction of adaptively rational actors. Listing 6 shows portions of the MAKE.OFFER module.

```
Event OUT.CHECK given reading.actor
        :
        If out.check.set(reading.actor) is not empty
                let best = f.out.check.set(reading.actor)
                let status(best) = .confirmed
        :

                activate an EXCHANGE giving best, reading.actor and to.node(best) now
        endif
END
```

Listing 5.

Event MAKE.OFFER given offer.actor and target.actor
:
Create an offer.made
Create an offer.recd
:
Let amountx(offer.made) = LEVEL.ASP(offer.actor)
Let amountx(offer.recd) = 24 - amountx(offer.made)
:
Print ...
:
END

Listing 6.

The ADAPT.LEVEL Module

This is a routine that simply lowers the level of aspiration by one unit. The rejections in IN.CHECK are communicated to the initiators of the offers, each of whom (in an instance of ADAPT.LEVEL) rationally adapts by lowering level of aspiration as little as seems necessary, that is, by one unit. Listing 7 shows the key statement of the ADAPT.LEVEL module.

The EXCHANGE Module

This is a routine that, acting on a confirmation generated by an OUT.CHECK, records an exchange, including the actors, the offer, and the time. It also updates the state of the active network, deleting the exchanging actors and testing for the end of the network process. Listing 8 presents parts of the EXCHANGE module.

This concludes our discussion of the modules. Before illustrating the outcomes generated by the simulation model consisting of these interrelated modules, we discuss one further model rule pertaining to the initial levels of aspiration. For each actor, this is defined by the smaller of 24 (the total number of points) and a term given by a nondecreasing function of the number of alternative exchange partners. In the program this function is given by $12 + d^2$, where d is the degree of the node, the number of its links. Consider a particular exchange network configuration such as that shown on the top of Table 2: a line with an odd number of nodes. For this 5-node line network, the initial levels of aspiration are given by: 13, 16, 16, 16, 13, for actors at positions 1 through 5, respectively. The first part of Table 2 shows a

Event ADAPT.LEVEL given rational.actor
:
Let LEVEL.ASP(rational.actor) = LEVEL.ASP(rational.actor) - 1
End

Listing 7.

Event EXCHANGE given offer, first.actor and second.actor
:
    Print 1 line with id.num(offer), first.actor, second.actor, amountx(offer) and 24-amountx(offer) thus
EXCHANGE (ID *** Time ***.**) ACTORS *** *** AMOUNTS ** **
    :
    Let activity.state(first.actor) = .inactive
    Let activity.state(second.actor) = .inactive
    :
END

Listing 8.

summary of 100 replications of the network exchange process. The second part shows a typical detailed output consisting of a sequence of offers (and, of course, implied counter-offers) culminating in two successive exchanges, the second of which terminates the particular replication. In this instance, the middle node (3) is excluded from exchange. This turns out to be emblematic of the general form of the outcome we see in the final summary over all 100 replications. This result indicates that, despite the high initial level of aspiration of this middle node—based on its perception of a choice between two exchange partners—this actor usually is excluded from exchanges: more than 85 percent of the exchanges of actor 2 are with actor 1 rather than actor 3 and a similar statement holds for actor 4 in relation to actors 5 and 3. Also, the average amounts earned shows a strong regularity: The actors in the even positions earn an average of about 13 points, the two endpoints an average of about 9 points, and the middle node an average of only 3 points.

What is happening here? People in the node 3 role are subjected to the ineluctable consequences of *the bearing of indirect ties* on their own fates: their exchange partners have their own alternatives, with respect to which they are well-positioned because these others have *no* alternative partners.

This is in agreement with the network analysis that is built into the "graph power index" or GPI developed by Markovsky, Willer, and Patton (1988). This index assigns a number to each node that reflects the ramifications of network structure for the bargaining power of each position in the network. The GPI treatment of the problem of generating exchanges in these networks has a comparative statics status: a hypothesis that the equilibrium outcomes will reflect the GPI indices of the nodes. It is not a consequence of a dynamic model. By contrast, given our simulation model, the figures are equilibrium outcomes of a dynamic model. Put in terms of micro-macro linkage, the GPI hypothesis relates macro givens to macro outcomes, while the DES simulation *generates* the macro outcomes described in general terms earlier when the Coleman micro-macro formulation was instantiated. We see now, however, a difference from Coleman's own exchange models, which are based on the general equilibrium theory of a market economy. The primary analytical goal of the latter is to solve for the exchange equilibrium in terms of given initial conditions. Process becomes secondary in analytical significance. By contrast, our

Table 2. Simulation of the 5-Node Line Network

1—2—3—4—5

Summary Averages Over 100 Replications

Actor	Avg Num Offers Sent	Avg Num Offers Recd	Avg Points	
1	4.73	3.67	9.42	
2	5.84	7.61	13.05	
3	5.93	4.36	3.16	
4	5.79	7.43	13.04	
5	4.38	3.60	9.33	

First Actor	Second Actor	Number of Exchanges	Average Per Exchange	
			First Actor	Second Actor
1	2	87	10.83	13.17
2	3	13	12.23	11.77
3	4	14	11.64	12.36
4	5	86	13.15	10.85

Example: A Single Generated Event History
Offer 1 At Time 6.59 From 1 To 2:13 11
Offer 2 At Time 8.84 From 2 To 1:16 8
Offer 3 At Time 9.66 From 5 To 4:13 11
Offer 4 At Time 13.96 From 4 To 5:16 8
Offer 5 At Time 14.75 From 5 To 4:12 12
Offer 6 At Time 15.35 From 2 To 1:16 8
Offer 7 At Time 15.70 From 5 To 4:12 12
Offer·8 At Time 16.25 From 5 To 4:12 12
Offer 9 At Time 17.26 From 4 To 5:15 9
Offer 10 At Time 19.66 From 4 To 5:15 9
Offer 11 At Time 21.33 From 3 To 2:16 8
Offer 12 At Time 22.50 From 5 To 4:11 13
Offer 13 At Time 23.90 From 4 To 5:14 10
Offer 14 At Time 25.36 From 2 To 3:16 8

Exchange (ID 13 Time 25.86) Actors 4 and 5: Amounts 14 10
Offer 15 At Time 26.64 From 2 To 3:16 8
Offer 16 At Time 27.63 From 1 To 2:12 12
Offer 17 At Time 30.43 From 1 To 2:12 12
Offer 18 At Time 31.04 From 2 To 1:15 9
Offer 19 At Time 32.65 From 1 To 2:11 13
Offer 20 At Time 33.21 From 2 To 1:14 10
Offer 21 At Time 33.97 From 2 To 3:14 10

Exchange (ID 20 Time 36.29) Actors 2 and 1: Amounts 14 10

Table 3. Nine Experimental Exchange Networks

Exchange Network	Simulation Node Numbers	Matrix	Structural Form

Row A:

Simulation Node Numbers:
```
1——2——3
    |
    4
```

Matrix:
$$\begin{pmatrix} 0 & 1 & 0 & 0 \\ 1 & 0 & 1 & 1 \\ 0 & 1 & 0 & 0 \\ 0 & 1 & 0 & 0 \end{pmatrix}$$

Structural Form:
```
A——B——A
    |
    A
```

Row B:

Simulation Node Numbers:
```
1—2—3—4
```

Matrix:
$$\begin{pmatrix} 0 & 1 & 0 & 0 \\ 1 & 0 & 1 & 0 \\ 0 & 1 & 0 & 1 \\ 0 & 0 & 1 & 0 \end{pmatrix}$$

Structural Form:
```
A—B—B—A
```

Row C:

Simulation Node Numbers:
```
1——2——3
    |
    4
    |
    5
```

Matrix:
$$\begin{pmatrix} 0 & 1 & 0 & 0 & 0 \\ 1 & 0 & 1 & 1 & 0 \\ 0 & 1 & 0 & 0 & 0 \\ 0 & 1 & 0 & 0 & 1 \\ 0 & 0 & 0 & 1 & 0 \end{pmatrix}$$

Structural Form:
```
A——B——A
    |
    C
    |
    D
```

Row D:

Simulation Node Numbers:
```
1——2——3——4——5
        |
        6
        |
        7
```

Matrix:
$$\begin{pmatrix} 0 & 1 & 0 & 0 & 0 & 0 & 0 \\ 1 & 0 & 1 & 0 & 0 & 0 & 0 \\ 0 & 1 & 0 & 1 & 0 & 1 & 0 \\ 0 & 0 & 1 & 0 & 1 & 0 & 0 \\ 0 & 0 & 0 & 1 & 0 & 0 & 0 \\ 0 & 0 & 1 & 0 & 0 & 0 & 1 \\ 0 & 0 & 0 & 0 & 0 & 1 & 0 \end{pmatrix}$$

Structural Form:
```
A——B——C——B——A
        |
        B
        |
        A
```

Row E:

Simulation Node Numbers:
```
1——2——3——4——5
        |
        6
```

Matrix:
$$\begin{pmatrix} 0 & 1 & 0 & 0 & 0 & 0 \\ 1 & 0 & 1 & 0 & 0 & 0 \\ 0 & 1 & 0 & 1 & 0 & 1 \\ 0 & 0 & 1 & 0 & 1 & 0 \\ 0 & 0 & 0 & 1 & 0 & 0 \\ 0 & 0 & 1 & 0 & 0 & 0 \end{pmatrix}$$

Structural Form:
```
A——B——C——B——A
        |
        D
```

Row F:

Simulation Node Numbers:
```
        3
1——2  <
        4
```

Matrix:
$$\begin{pmatrix} 0 & 1 & 0 & 0 \\ 1 & 0 & 1 & 1 \\ 0 & 1 & 0 & 1 \\ 0 & 1 & 1 & 0 \end{pmatrix}$$

Structural Form:
```
         D
A——B  <
         D
```

Row G:

Simulation Node Numbers:
```
1        4
 >  3  <
2        5
```

Matrix:
$$\begin{pmatrix} 0 & 1 & 1 & 0 & 0 \\ 1 & 0 & 1 & 0 & 0 \\ 1 & 1 & 0 & 1 & 1 \\ 0 & 0 & 1 & 0 & 1 \\ 0 & 0 & 1 & 1 & 0 \end{pmatrix}$$

Structural Form:
```
A        A
 >  B  <
A        A
```

Row H:

Simulation Node Numbers:
```
1         4
 >  3  <
2         5——6
```

Matrix:
$$\begin{pmatrix} 0 & 1 & 1 & 0 & 0 & 0 \\ 1 & 0 & 1 & 0 & 0 & 0 \\ 1 & 1 & 0 & 1 & 1 & 0 \\ 0 & 0 & 1 & 0 & 1 & 0 \\ 0 & 0 & 1 & 1 & 0 & 1 \\ 0 & 0 & 0 & 0 & 1 & 0 \end{pmatrix}$$

Structural Form:
```
A         C
 >  B  <
A         D——E
```

Row I:

Simulation Node Numbers:
```
1——2——3——5——6
        \ /
         4
```

Matrix:
$$\begin{pmatrix} 0 & 1 & 0 & 0 & 0 & 0 \\ 1 & 0 & 1 & 0 & 0 & 0 \\ 0 & 1 & 0 & 1 & 1 & 0 \\ 0 & 0 & 1 & 0 & 1 & 0 \\ 0 & 0 & 1 & 1 & 0 & 1 \\ 0 & 0 & 0 & 0 & 1 & 0 \end{pmatrix}$$

Structural Form:
```
A——B——C——E——F
        \ /
         D
```

approach gives primacy to process models. Not only the equilibrium results, but the event history of an exchange network process is generated in our model.

Table 3 shows nine networks recently the subject of extensive theoretical and empirical study (Willer 1992). In Table 4 are displayed the summary results generated by our DES model for each of these networks. The outcomes are in good qualitative agreement with experimental results (see Table 1 in Skvoretz and Fararo 1992a). By node positions, the amounts earned in exchange (rounded in some instances) are presented in Table 5.

One patterned departure (appearing in networks A, C, and D) is that actors in stronger positions tend to earn more in experiments than is predicted by the model. This is consistent with the dynamics of offers as well. Generally, the real offer event histories begin with more asymmetric offers by those in power positions than our model generates. But to some extent these departures could represent a difference in knowledge states in the experiments and in the model. The experimenters often

Table 4. Output of Network Program for Nine Experimental Networks Summary Tables

Summary for Network A: Averages Over 100 Replications				
Actor	Avg Num Offers Sent	Avg Num Offers Recd	Avg Points	
1	5.06	3.49	3.52	
2	9.54	14.60	15.12	
3	4.71	3.03	2.73	
4	4.83	3.02	2.63	
				Average Per Exchange
First Actor	Second Actor	Number of Exchanges	First Actor	Second Actor
1	2	40	8.80	15.20
2	3	30	14.90	9.10
2	4	30	15.23	8.77
Summary For Network B: Averages Over 100 Replications				
Actor	Avg Num Offers Sent	Avg Num Offers Recd	Avg Points	
1	5.12	4.00	10.33	
2	5.53	6.64	13.19	
3	5.58	6.42	13.17	
4	4.89	4.06	10.35	
				Average Per Exchange
First Actor	Second Actor	Number of Exchanges	First Actor	Second Actor
1	2	96	10.76	13.24
2	3	4	12.00	12.00
3	4	96	13.22	10.78

(continued)

Table 4. (continued)

Summary for Network C: Averages Over 100 Replications

Actor	Avg Num Offers Sent	Avg Num Offers Recd	Avg Points
1	5.56	4.03	3.81
2	9.38	13.04	15.12
3	5.76	4.10	5.07
4	4.73	5.75	13.15
5	4.50	4.01	10.85

First Actor	Second Actor	Number of Exchanges	Average Per Exchange First Actor	Second Actor
1	2	43	8.86	15.14
2	3	57	15.11	8.89
4	5	100	13.15	10.85

Summary For Network D: Averages Over 100 Replications

Actor	Avg Num Offers Sent	Avg Num Offers Recd	Avg Points
1	4.82	3.45	10.92
2	5.56	7.21	13.08
3	8.55	7.37	.67
4	6.12	7.64	13.00
5	4.71	3.59	10.62
6	6.32	7.78	13.08
7	4.55	3.59	10.63

First Actor	Second Actor	Number of Exchanges	Average Per Exchange First Actor	Second Actor
1	2	100	10.92	13.08
3	4	3	12.67	11.33
3	6	2	14.50	9.50
4	5	97	13.05	10.95
6	7	98	13.15	10.85

Summary for Network E: Averages Over 100 Replications

Actor	Avg Num Offers Sent	Avg Num Offers Recd	Avg Points
1	4.71	3.63	10.89
2	5.51	6.41	13.11
3	10.10	11.55	15.01
4	5.37	6.23	13.16
5	4.66	3.62	10.71
6	7.92	6.83	8.88

First Actor	Second Actor	Number of Exchanges	Average Per Exchange First Actor	Second Actor
1	2	100	10.89	13.11
3	4	1	13.00	11.00
3	6	99	15.03	8.97
4	5	99	13.18	10.82

(continued)

51

Table 4. (continued)

Summary For Network F: Averages Over 100 Replications

Actor	Avg Num Offers Sent	Avg Num Offers Recd	Avg Points
1	8.15	6.34	8.80
2	10.19	12.83	14.90
3	7.53	6.85	11.61
4	6.74	6.59	11.97

			Average Per Exchange	
First Actor	Second Actor	Number of Exchanges	First Actor	Second Actor
1	2	97	9.07	14.93
2	3	1	14.00	10.00
2	4	2	14.00	10.00
3	4	97	11.87	12.13

Summary for Network G: Averages Over 100 Replications

Actor	Avg Num Offers Sent	Avg Num Offers Recd	Avg Points
1	7.26	7.30	11.48
2	7.72	7.25	11.84
3	10.88	10.84	1.38
4	7.66	7.49	11.56
5	6.55	7.19	11.74

			Average Per Exchange	
First Actor	Second Actor	Number of Exchanges	First Actor	Second Actor
1	2	95	11.96	12.04
1	3	1	12.00	12.00
2	3	4	10.00	14.00
3	4	3	14.67	9.33
3	5	2	13.00	11.00
4	5	95	11.87	12.13

Summary For Network H: Averages Over 100 Replications

Actor	Avg Num Offers Sent	Avg Num Offers Recd	Avg Points
1	7.54	7.05	11.46
2	7.14	7.04	11.92
3	13.84	14.67	12.62
4	11.22	9.86	9.21
5	10.07	13.31	14.74
6	7.41	5.29	7.73

			Average Per Exchange	
First Actor	Second Actor	Number of Exchanges	First Actor	Second Actor
1	2	96	11.77	12.23
1	3	2	8.00	16.00
2	3	2	9.00	15.00
3	4	82	14.63	9.37
4	5	15	10.20	13.80
5	6	85	14.91	9.09

(continued)

Table 4. (continued)

Summary For Network I: Averages Over 100 Replications

Actor	Avg Num Offers Sent	Avg Num Offers Recd	Avg Points
1	5.15	3.42	10.62
2	5.41	7.41	13.12
3	10.91	10.50	12.09
4	9.49	8.60	10.00
5	10.26	12.43	14.88
6	7.11	5.97	7.93

			Average Per Exchange	
First Actor	Second Actor	Number of Exchanges	First Actor	Second Actor
1	2	98	10.84	13.16
2	3	2	11.00	13.00
3	4	86	13.76	10.24
4	5	12	9.92	14.08
5	6	88	14.99	9.01

tell the subjects the shape of the network and their position in it, thus providing a global map of the system that actors in real networks are likely to desire but not possess.

However, it is also true that our model includes some arbitrary elements that might contribute to the discrepancy. Two such elements are the arbitrary character of the rule that yields the initial aspiration levels and the rather arbitrary selection of time delays between certain events. Such arbitrary elements that tend to creep into models generally, and simulation models in particular, detract from their theoretical significance.

In conclusion, the DES model embodies a sound micro-macro dynamic logic; it generates plausible offer event histories; it captures the main effects of network constraints on exchange outcomes particularly noticeable in terms of exclusions from exchanges; and it is in qualitative agreement with experimental results on varying networks. Despite these positive aspects, it needs further development to enhance its theoretical value.

Table 5. Comparison of DES Model with Experimental Results

Network	DES Model	Experimental Results
A	B15-A9	B20-A4
B	B13-A11 and B12-B12	B12.5-A11.5 and B12-B12
C	B15-A9 and C13-D11	B19-A5 and C12.2-D11.8
D	B13-A11 (C excluded almost always)	B18-A6
F	B15-A9 and D12-D12 [B-D rarely exchange]	B14-A10 and D12-D12
G	A12-A12 [B usually excluded]	A12-A12

This section has been a short introduction to the implementation of DES in the Simscript language as well as an extended discussion of one DES theoretical model We have seen how the key constructs, entities with attributes, are implemented in the theoretical model and in the simulation language in declarations and in modules that represent recurrent events and processes. Simscript has a natural language orientation and introduces several built-in elements that simplify the programming of a simulation model, features that enhance its attractiveness as one way of implementing DES simulation methodology.

THEORETICAL MODEL BUILDING WITH DES MODELS

The previous section discussed an exchange network model to explain the Simscrip implementation of the logic of discrete event simulation model building. The following discussion extends this work of showing how DES methodology and its Simscript implementation can function in theoretical research programs. There is no suggestion that the types of sociological problems treated earlier and in this section exhaust the possibilities: far from it, the claim is that wherever one is seeking to represent process, the approach is a relevant and important theoretical method.

Like the exchange network model, the next example exhibits a logic of "theoretical models in progress." This usually means starting with initial simplifications and then adding complications in successive revisions. In programming terms, any one theoretical model becomes successively embodied in a series of programs, the later programs correcting and extending the earlier. In tandem with this process the theoretical model is corrected or extended and does not remain a static given At any one point in this series of developments, a simulation model is both a theoretical model and a program. There is really never a last program in the series, only a place of rest or termination through exhaustion of the creative possibilities or diversion into work on other such projects. Any one series is responsive both to internal logical critical development and to revisions motivated by comparisons of the implications of the models (through study of the reported results of the programs) with empirical data or known empirical regularities to be accounted for

E-State Structuralism and Dominance Structures

The theoretical problem concerns the dynamics of interlocking relational actions of actors in which each actor has a relational orientation state, termed an E-state, toward particular or generalized others. Patterns of such E-states and how they evolve are the focus of attention. Such E-states are abstract conceptions of expectation states. Involved here is what has been called "E-state structuralism" (Fararo and Skvoretz 1986). The state of a social network is represented as a set of such E-states, one for each ordered pair of actors in the network. A social structure is an equivalence class of such network states such that the class constitutes a stable equilibrium of a structure-formation process over the network.[8] Thus, the network is dynamic and any stability of form it possesses is conceptualized as an outcome of process.

The micro-macro logic described earlier is built into this E-state structuralism. Recall that, in our formulation, the macro level is characterized by the state of a social system, earlier described as a structure. In the models to be described, this structure is now not a fixed parameter but a network state that is evolving in time and may, under certain conditions, come to possess a certain stability of form so as to constitute a social structure. Recall, also, that in our formulation the micro level is characterized by the situation of an actor and the actor's action in that situation. In the present models, the other actors in the network are the key situational objects for the actor and the actor's action is a function of the current E-state of the actor toward others.

The main difference, in micro-macro logic, between the exchange network model and the present models is that adaptive rationality is not explicitly represented in the E-state models worked out so far. It is likely that this will be done in the future, but for the present the principle that governs action in situations is as follows.

An E-state is related to action in that: (a) such a relational state may arise due to actor-generated events in the nexus of interactions of various actors, and (b) once such an E-state exists toward another actor, actions depend on the state. This is the logic of what has been called "state-organizing" processes (Berger, Wagner, and Zelditch 1985) in the expectation states research program that forms part of the background for E-state structuralism as a theoretical method. In the theoretical models to be described, E-states play a role analogous to the levels of aspiration in the exchange network model.

Several theoretical model-families are being developed that relate to the problem of generating dominance structures. These types of social structures have been the subject of extensive field work in sociobiology (see Wilson 1975, Chapter 13). There also have been controlled experiments involving animals (see especially Chase 1982). The ultimate aim of the research program is to capture the basic empirical regularities found in this work as outcomes of dynamic processes of interaction formulated in terms of the E-state structuralist logic. Because we are thinking of models of dominance structures in this social biological context, henceforth we sometimes write "animal" for the more abstract term "actor" in describing the models.

The basic process rule in both types of models is this: if actor A is in a deferential E-state toward an animal B, then A does not attack B. On the other hand, if no E-state orientation exists or if the animal has a dominance orientation toward the other, an attack may occur. The concept of a dominance *relation* is defined as a pair of complementary E-states: A is in a dominance relation to B if and only if A has a dominance E-state toward an actor B and B has a deferential E-state toward A. In such a relation, the superiority-inferiority is established through the complementary states: it is an emergent asymmetric social relation. Because of the rule that deference implies no attacks, it tends to be a stable state in its own right. A dominance structure is a pattern of E-state orientations analyzed with respect to the extent to which it exhibits dominance relations, transitivity of such relations, and the like. The fundamental idea is to show how these E-state orientations evolve through social interactions, yielding a variety of paths of change of the dominance patterns until some unique equilibrium structure or some fixed probability distribution over structures is obtained. In the first family of models, which will be termed E-state1 models, the complementarity of E-states is not taken as problematic. Instead, the event that animal A develops a certain orientation to B happens instantaneously with the event that B develops the complementary orientation. For instance, with a certain probability, given no relation has yet formed, animal A attacks B. With another probability, A forms a dominance relational orientation toward B as a consequence of this attack. In E-state1 models, at the same time, B forms a deferential orientation toward A.

In the second family of models, called E-state2 models, a complementary orientation of B toward A does not necessarily form—not only not instantaneously, but ever. The result is a much more complex set of possible outcomes. E-state1 models have been developed further because we began with the simpler type of model. Most recently, a third family of models is being explored in which the main mechanisms and outcomes of these two models are embedded as special cases. For the purpose of this paper, we restrict our discussion to the E-state1 and E-state2 models. All three families of models are discussed in great detail in Fararo, Skvoretz, and Kosaka (1994).

E-State1 Dominance Structure Formation Models

The theoretical problem is to formulate an explicit process such that we can generate the typical forms of dominance structures found in observational and experimental studies of animal groups. A mathematical model embodying a bystander mechanism is our first version of the E-state1 family of models responsive to this problem but with consequences derived for only three or four animals in the initial formulation (Fararo and Skvoretz 1986, 1988).

The DES simulation task was to embody the E-state1 model in a computer program so as to explore the properties of this model beyond those cases where analytical techniques could derive formulas. The simulation model is effective in generating tables of outcome dominance structure properties as a function of varying parameters relating to, for example, a basic mechanism in the model that deals with bystanders. For instance, one derived quantity is the probability that the emergent structure is a complete hierarchy. As the size of the group increases, this outcome becomes less and less likely, as other and more complex structural forms arise.

The DES simulation model is of some interest in another respect. The detailed action-event histories such as the one shown in Table 6, model *process parallelism*:

Table 6. Simulation of Emergence of Dominance Structures

For One Group of Size 6
Parameters: Pi = .50, Theta = .90

Attack Number 1: 2 Attacks 3 Starting At 0
  2 Dominates 3
Number Of Bystanders Is 4
  1 Is A Bystander: 2 Dominates 1
  4 Is A Bystander: 4 Dominates 3, 2 Dominates 4
  5 Is A Bystander: 5 Dominates 3, 2 Dominates 5
  6 Is A Bystander: 2 Dominates 6
Attack Number 1 Ends At .730
Attack Number 3: 2 Attacks 6 Starting At 3.136
Number Of Bystanders Is 0
Attack Number 3 Ends At 3.178
Attack Number 2: 5 Attacks 1 Starting At 2.517
Number Of Bystanders Is 3
  3 Is A Bystander: 3 Dominates 1
  4 Is A Bystander: 4 Dominates 1, 5 Dominates 4
Attack Number 2 Ends At 3.525
Attack Number 4: 2 Attacks 5 Starting At 6.106
Number Of Bystanders Is 2
  1 Is A Bystander: 1 Dominates 5
  4 Is A Bystander
Attack Number 4 Ends At 6.401
Attack Number 5: 5 Attacks 6 Starting At 8.404
  5 Dominates 6
Number Of Bystanders Is 4
  1 Is A Bystander: 1 Dominates 6
  2 Is A Bystander
  3 Is A Bystander: 3 Dominates 6
  4 Is A Bystander: 4 Dominates 6
Attack Number 5 Ends At 9.959
Time To Complete Group 1 Dominance Structure = 12.501 (Minutes)
  Pairs Such That First Dominates Second Are:
    1 5, 1 6
    2 1, 2 3, 2 4, 2 5, 2 6
    3 1, 3 6
    4 1, 4 3, 4 6
    5 3, 5 4, 5 6

The Transitivity Of The Dominance Relation Is .90

attacks can occur concurrently, in parallel, so that an aggressive encounter takes some time and may overlap with other attacks involving other animals. This is possible because each encounter is represented as a Simscript process that takes some time rather than as an instantaneous event. In a reasonably large assemblage of animals, only a subset of them become bystanders to a given attack. When an attack begins, a set of bystanders assemble to observe it. But before it ends, some of these bystanders may be diverted from observing this attack by virtue of becoming attackers or attackees themselves. Thus another encounter starts before a current one is completed, illustrating parallelism.

One can get a feel for the translation of ideas about the social phenomena involved into Simscript from the following fragments of the program. There are two processes, called GENERATOR and ATTACK. Because processes are routines, they can have arguments. For instance, ATTACK instances are numbered. There are two substantive parameters, the probability, called pi here, that an attack leads to the formation of a dominance relation between attacker and attackee and the probability, called theta here, that a bystander-to-attacker (attackee) relation forms by observation of the attack. The basic idea is that this event mirrors the direction-ality of the attack: a bystander (not currently in dominance relation to the relevant other) forms a deference orientation to the attacker (and, in another event, a dominance orientation toward the attackee). It is this bystander mechanism that is the fundamental mechanism of this model. There is a Simscript set conceptualized as a system property and called "the pool," that is a dynamically varying list of eligibles for a role in a generated attack. For instance, if animal A is involved in some attack k (the attack.number here) we postulate that A is not in the pool for a new attack as either attacker, attackee, or bystander until attack k is over. A more complex model would permit an animal to become the attackee of a second attack while still involved as attacker or attackee in a given attack. So in this simpler model only bystanders to an ongoing attack may be diverted into a new attack situation as attacker or attackee. Considerations of simplicity, as in all model building, together with the concept of gradual development of a sequential model family, are at work here.

The matrix D declared here begins as an all-zero matrix: no dominance relation-ships exist. When an attack creates a transition of E-state so that now animal A dominates animal B, the corresponding entry in the matrix is set to 1. This matrix evolves toward an equilibrium through the generation of attacks that are occasions of changes of state of orientation of animals toward each other. Excerpts from the various modules follow in Listing 9.

In this model, we automatically tally the average proportion of transitive triads and, when control returns to main after all replications are done, we print out a table of results (not displayed here). For each of a selected set of values of the key parameters (group size, pi, theta), the entry is the average transitivity under those parameter conditions. In this way, we are doing the equivalent of a logical derivation in a mathematical model: such a table is an approximation to a theorem about how the equilibrium outcome of a process depends on its parametric conditions, a comparative statics theorem.[9] To the extent that the comparative statics theorem (table) tends to agree with observations of real animal groups, these observations are explained. In other words, observations of group structures (in equilibrium, under a variety of conditions) correspond to derived outcomes of the simulation model and the mechanisms of the model explain the observations.

```
PREAMBLE
        Processes
                Include GENERATOR
                Every ATTACK has an attack.number
        Permanent entities
                Every animal may belong to the pool
        The system owns the pool
        Define pi, theta, prop.transitive,... as real variables
        Define D as an integer, 2-dimensional array
        :
        Tally avg.prop.transitive as the average of prop.transitive
END

MAIN
        :
        Create every animal(N)
        :
        Activate a generator now
        Start simulation
        :
END

Process GENERATOR
        Until complete = yes
                Do
                        Add 1 to attack.no "becomes the new attack.number
                        Activate an attack giving attack.no now
                        Wait exponential.f(2,2) minutes
                Loop
END

Process ATTACK given attack.number
        : (Below "animal" refers to a bystander)
        If random.f(8) LE theta and D(attackee,animal) = 0 and D(animal,attackee) = 0
                let D(animal,attackee) = 1 "transition of E-state
                If printguide = "yes" ((want detailed attack output))
                        print 1 line with animal and attackee thus
** DOMINATES **
        Endif
        :
END
```

Listing 9. E-state1 model

In terms of micro-macro logic of Figure 1, framed in the Coleman-type manner, what we have is the following. There is an initial macro state characterized by an assemblage of animals without a dominance structure. This implies null E-states throughout the system. This macro fact determines the micro-level situation of any animal as unstable, that is, as not consisting of E-states. Acts of aggression occur and these yield certain E-states that constitute dominance relations, a particular micro-macro transition. Eventually, the complete network of dominance relations is thereby constructed. The initial macro-level fact of no dominance structure has been transformed, over time, into a novel macro-level fact, a dominance structure. The DES model provides the dynamical model of how actions in situations produce this transformation.

Our purely theoretical knowledge consists of this: we understand the logical structure of our theoretical models and their logical implications. In simulation, the logical structure is embedded in declarations of types of entities and their attributes and in the postulation of processes embodying mechanisms and the logical implications are the outcomes generated under varying parametric conditions, usually over a sufficiently large set of replications to accommodate the existence of some randomness in the processes involved.

As it turns out, the tables we obtain show not only the importance of the bystander mechanism—as discussed in empirical terms by Chase (1982) and in terms of the logic of E-state structuralism by Fararo and Skvoretz (1986)—but also a systematic size effect. The transitivity tends to decline as the size increases, at least for the sizes from 3 to 6 subjected to large-scale replications. Similarly, the probability of a perfect hierarchy which ranges from about .75 to 1 (as pi and theta increase) for groups of size 3, for larger groups shows quite different approximate ranges:

size 4: .38 to .81

size 5: .12 to .67

size 6: .03 to .47

E-State2 Dominance Structure Formation Models

The theoretical problem here is to remove a simplifying assumption made in the E-state1 model family: that if E-states are formed, they simultaneously form in a complementary manner.

Thus far only the very first theoretical model has been created and studied by DES methodology, using Simscript. The model restricts itself initially to the triad to work out the basic logic. The state of a triad is described by a configuration, consisting of the values of each of six variables, an E-state for each animal toward every other: a dominance orientation, a neutral orientation or a deferential orientation. Therefore, there are 729 possible states of a triad. In this model we allow a complexity not treated in the E-state1 family: not only are E-states formed without automatic complementariness, but also we allow transitions back from deference to neutral or from dominance to neutral. Thus, at a given time, when animal B is deferential to A but B observes an attack of C on A, the attack may be the occasion for B to shift back to neutral, then later, if A is attacked again with B as a bystander, B may switch to a dominance orientation to A. Thus, repeated attacks on an animal can have the effect of reducing its earlier dominance status. This is possible because to have a dominance orientation to another does not imply the complementary deference necessarily and therefore the former animal might still be attacked, contrary to its unilateral dominance orientation. Recall that the fundamental axiom is that only when an animal is in a deferential orientation to another will that animal not attack the other. As a consequence of this problematic aspect of complementariness, a variety of distinct relations between animals can arise and be maintained in equilibrium.

As a first effort, we describe these relations by extending the well-known roster of structural types of triads in network theory, called MAN (Holland and Leinhardt 1977) and reinterpret them for this more dynamic context. MAN stands for mutual, asymmetric, and null relations implied by a single "choice" relation in sociometry. We add a conflict relation, yielding a roster of 29 triad types called MANC.

We define the MANC relations in terms of E-states as follows (where we also show the implied observable behavior):

Mutuality:
 E-states: Each is in a deferential state toward other.
 Behavior: Neither attacks the other.
Asymmetry:
 E-states: Complementary dominance-deference orientations.
 Behavior: B never attacks A, but A may attack B.
Null:
 E-states: At least one is neutral toward the other.
 Behavior: Either one-way or two-way attacks may occur.
Conflict:
 E-states: Each has a dominance orientation toward other.
 Behavior: Attacks upon each other.

The problem is to show how the equilibrium triad type or, in other cases, the probability distribution over a set of such types, emerges and depends upon the parametric conditions.

Since the problem is too difficult to handle mathematically, it is treated in terms of discrete event simulation methodology and embedded in a Simscript program. The program uses a compound permanent entity (a directed tie between two animals) with an E-state as an attribute that will exhibit change as attack process instances are generated: attacker, attackee, and the bystander all may exhibit change of E-state. Because of the discrete event character of the simulation, there is not just a sequence of attacks along an ordinal time axis, as in a discrete time model. Instead, the time domain is continuous and attacks are generated as a stochastic process, each attack having an associated time of occurrence. The program does not look much different from the E-state1 program, so fragments of it will not be shown here.

What is the logical implication of the E-state2 idea that complementariness is problematic and that E-states may revert to neutral and then shift to the opposite orientation, depending on the events that occur? A quite surprising result of this simulation model is that under a very wide set of parametric conditions what emerges is not a transitive hierarchy but a coalition of two against one: the two mutually defer and both dominate the third, who defers to each of them, an equilibrium state. Like theories expressed in mathematical form and then analyzed logically to derive theorems, simulation models can have surprising properties. Indeed, "surprise" is one of the elements of assessment of a model (Lave and March 1975): it has logical consequences that are not obvious from its defining assumptions.

This model is also reported in Fararo, Skvoretz, and Kosaka (1994) and, as mentioned earlier, a third type of model is constructed that captures the E-State1 and E-State2 types as special cases.

CONCLUDING DISCUSSION

In this paper, we have focused on the logic of discrete event simulation as a theoretical method. To begin, it was pointed out that simulation is becoming recognized as an important tool for theory building but that there was a need for careful explications of the theoretical logic of simulation with real examples of its use in theoretical research programs. The initial discussion dealt with the logic of dynamic models, discussed in terms of choices from six menus: alternatives for the specification of the state and parameter spaces (discrete or continuous), the time domain (discrete or continuous), the timing of events (regular, incessant, irregular), the generator (deterministic or stochastic), and the postulational basis (equations or transition rules). A simulation was defined as a dynamic model implemented on a computer. Various types of simulation were noted with a basic contrast existing between continuous simulations and discrete event simulations.

A discrete event simulation model in which there is a system of interdependent entities and their activities is described in terms of the following: *categorical states*, a *continuous time domain*, with events occurring at *irregular times*, and a postulational basis of *transition rules* among the categorical states. An event is a change of state and in such models, events occur at separated and usually stochastically generated time points, implying a series of discontinuous changes of state.

Then, discrete event methodology was described in terms of the idea that the basic elements are entities and their attributes. Entities may be permanent or temporary. They may be not only singular units but compounds or relations, such as ties treated as entities. Attributes may be fixed for the duration of a dynamic analysis or they may be time-variable, in which case they are state variables. State variables change when events occur. In Simscript, these and other ideas were implemented in a simulation programming language. It was shown how Simscript automates much of the work of simulation programming, since its built-in timing routine handles most of the details that the user does not have to worry about. The focus then becomes the substance of the theoretical model translated into Simscript language.

As a first introduction to discrete event methodology via the Simscript language implementation, a simulation model addressing the production of phenomena in exchange networks was presented. Because of the length of the program called NETWORK, only fragments of each program module were shown. The entities were identified, certain states indicated and the way in which states changed was addressed in terms of the various events. The NETWORK program embodies a model and this model was studied in terms of a variety of exchange networks, comparing the generated outcomes with experimental data. The qualitative agreement is good, although not perfect, encouraging continued work and theoretical refinements.

In the prior section, the logic of the NETWORK model was reversed: instead of showing how a certain phenomenon varies with the structure of a network, the problem was to show how a network structure emerges from initial conditions in which it does not exist. A theoretical method called E-state structuralism was used, now blended with the discrete event simulation methodology. Two families of models that treat the problem of generating dominance structures were discussed. In the E-state1 model family, it was shown that a simulation model, like a mathematical model presented in analytical form, has certain definite logical consequences that are discovered, not by deduction, but by repetitions of the simulation to discover implied regularities. In particular, the variation in transitivity by group size was noted as one such logical consequence, functioning much like a theorem in the analytical study of theoretical models. In the E-state2 model family, distinguished by making complementary E-states problematic, this point was made again with the additional point that a rather surprising theorem-like regularity emerged: the two-against-one structure was the common outcome, as contrasted with the hierarchy generated by the typical E-state1 model for the triad. In turn, these two quite different structural outcomes have prompted another research problem leading to a third type of model that captures these two as special cases.

In both types of role of social structure, as given parameter (in the exchange network model) and as emergent outcome of interaction (in the E-state models), we employed a micro-macro linkage approach which agrees with Coleman's (1990) general formulation of micro-macro logic while more strongly emphasizing the explicit modeling of processes that generate the macro phenomena of interest.

It is extremely important to remember that a simulation analysis, like a logico-mathematical analysis aimed at proving propositions, is getting at the necessary logical consequences of a set of assumptions. In simulation methodology, these assumptions are built into the declarations of entities and into the specification of routines that embody events and processes that create changes of state.[10] The utilization of discrete event simulation in sociology is just beginning. We are embarked on a collective learning process concerning its importance as a tool for theoreticians. This paper has tried to convey a sense of this importance by providing specific examples of theoretical models and research programs employing DES methodology along with other methods, especially an explicit dynamic implementation of micro-macro linkage.

A final point may be noted. In graduate education in sociology, we are very effective in teaching people how to analyze data. What we are deficient at doing is teaching people how to "think process." We need to teach our students how to construct models that generate social action episodes as well as summary numerical data. It is our hope that the spread and teaching of DES methodology and programming tools can be one vehicle for upgrading our teaching and research on the explanatory analysis of social action processes.

NOTES

1. At the same time, Hanneman's continuous simulation focus can be *embedded* within our approach. For instance, the language SIMSCRIPT enables the construction of continuous simulation, discrete simulation, and even *combined* simulation models. Thus, this approach allows the incorporation of both the sorts of formulations Hanneman elucidates and those that we will be stressing in our discussion of discrete event simulation methodology.

2. Hanneman's continuous simulation models are different in three basic ways: state description, his models involve continuous variables; timing of events, his models involve incessant change; and postulational basis, his models involve postulated equations.

3. Actually, even if the units arriving and departing were continuous variables, as long as the change of state occur only at discrete, that is, separated, instants, the process would still be a discrete event process.

4. See Chapter 2 of Fararo (1989) for an extended discussion of studying the implications of a theoretical model. For an extended example, see Kosaka and Fararo (1991).

5. The most important new development in software methodology involves object-oriented software design and programming. The new language C++ implements this methodology. Our current research involves embedding DES within C++ to achieve greater flexibility in model building than the special simulation language environments permit. Our intention is to report on this development at a later time.

6. The software and the manuals are available from CACI Products Company, LaJolla, California. The same company published the book, *Building Simulation Models in Simscript II.5* by Edward C. Russell. Although the examples in this book are in the operations research domain, the reader can use them to gain insights into Simscript implementations of discrete sociological process models.

7. Out listings are usually portions of the complete modules. The entire program listing is available on request.

8. Network states in the same class are both graph-isomorphic and homogenous in transitions of state, for example, if state A in class I goes to a state in class II, then any other state in class I also goes to class II, so that we can lump the states in the terminology of Markov chain theory.

9. See Fararo (1989, Chapter 2) for a general discussion of types of theorems.

10. There are at least two important features of Simscript that make it useful for developing theoretical models in sociology. The first already has been discussed and may be repeated for emphasis: it has an English-like syntax, making it more user-friendly for non-professional programmers. The second feature has not been mentioned yet. Simscript also incorporates continuous simulation processes within its language capabilities. The result is that one can create combined discrete-

continuous simulation models. This means not just that some attributes of some entities are continuous. It means that some of them change continuously. For example, in an ecosystem, the usual mathematical model treats the populations as continuous variables undergoing continuous change, as in familiar predator-prey models. In a simulation of such an ecosystem one might want to introduce the idea of event-driven changes in parameters of the ecosystem as an adjunct to the continuous process. Then the complete Simscript model would contain a representation of a recurrent event (or process) which, when it occurs at separated times, implies a discontinuous shift in one or more parameters of the ecosystem, which in turn would effect the continuous-change process in state space. Put another way, in such a model, the state space is the scene of standard continuous simulation, while the parameter space is the scene of discrete event simulation and the two are combined in one model. It is easy to imagine more sociological examples of a similar kind, with the events in parameter space perhaps triggered by certain thresholds for continuously variable phenomena in state space in a feedback loop.

REFERENCES

Alexander, J. C., B. Giesen, R. Munch, and N. J. Smelser, eds. 1987. *The Micro-Macro Link*. Berkeley: University of California Press.

Asch, S. 1951. "Effects of Group Pressure upon the Modification and Distortion of Judgment." Pp. 177–190 in *Groups, Leadership and Men*, edited by H. Guetzkow. Pittsburgh, PA: Carnegie Press.

Bainbridge, W. S. 1986. *Sociological Laboratory: Computer Simulations for Learning Sociology*. Belmont, CA: Wadsworth.

Berger, J., D. G. Wagner, and M. Zelditch, Jr. 1985. "Introduction: Expectation States Theory: Review and Assessment." In *Status, Rewards, and Influence*, edited by J. Berger and M. Zelditch, Jr. San Francisco: Jossey-Bass.

Carley, K. 1991. "A Theory of Group Stability." *American Sociological Review* 56:331–354.

Chase, I. 1982. "Dynamics of Hierarchy Formation: The Sequential Development of Dominance Relationships." *Behavior* 80:218–240.

Cohen, B. P. 1963. *Conflict and Conformity: A Probability Model and Its Application*. Cambridge, MA: MIT Press.

Cohen, B. P. 1989. *Developing Sociological Knowledge: Theory and Method*. Chicago: Nelson Hall.

Coleman, J. S. 1964. *An Introduction to Mathematical Sociology*. New York: Free Press.

———. 1990. *Foundations of Social Theory*. Cambridge, MA: Harvard University Press.

Collins, R. 1988. *Theoretical Sociology*. San Diego, CA: Harcourt Brace Jovanovich.

Cook, K. S., R. M. Emerson, M. R. Gillmore, and T. Yamagishi. 1983. "The Distribution of Power in Exchange Networks: Theory and Experimental Results." *American Journal of Sociology* 89:275–305.

Evans, J. B. 1988. *Structures of Discrete Event Simulation*. New York: Wiley.

Fararo, T. J. 1973. *Mathematical Sociology*. New York: Wiley. (Reprinted by Krieger, Melbourne, FL, 1978).

———. 1989. *The Meaning of General Theoretical Sociology*. ASA Rose Monograph. Cambridge, MA: Cambridge University Press.

Fararo, T. J. and J. Skvoretz. 1986. "E-State Structuralism: A Theoretical Method." *American Sociological Review* 51:591–602.

———. 1988. "Dynamics of the Formation of Stable Dominance Structures." In *Status Generalization*, edited by M. Webster and M. Foschi. Palo Alto, CA: Stanford University Press.

Fararo, T. J., J. Skvoretz, and K. Kosaka. 1994. "Advances in E-State Structuralism: Further Studies in Dominance Structure Formation." *Social Networks* 16: 233–265.

Feinberg, W. E. and N. R. Johnson. 1990. "Radical Leaders, Moderate Followers: Effects of Alternative Strategies on Achieving Consensus for Action in Simulated Crowds." *Journal of Mathematical Sociology* 15:91–115.

Freese, L., ed. 1980. *Theoretical Methods in Sociology: Seven Essays.* Pittsburgh, PA: University of Pittsburgh Press.

Hanneman, R. A. 1988. *Computer-Assisted Theory Building: Modeling Dynamic Social Systems.* Newbury Park, CA: Sage.

Heise, D. 1986. "Modeling Symbolic Interaction." Pp. 291–309 in *Approaches to Social Theory*, edited by S. Lindenberg, J. S. Coleman, and S. Nowak. New York: Russell Sage Foundation.

Holland, P. W. and S. Leinhardt. 1977. "A Method for Detecting Structure in Sociometric Data." In *Social Networks: A Developing Paradigm*, edited by S. Leinhardt. New York: Academic Press.

Hummon, N. P., ed. 1990a. *Computer Simulation in Sociology* (Special Issue of the *Journal of Mathematical Sociology*). New York: Gordon and Breach.

———. 1990b. "Organizational Structures and Network Processes." Pp. 149–161 in *Computer Simulation in Sociology* (Special Issue of *Journal of Mathematical Sociology*), edited by N. P. Hummon. New York: Gordon and Breach.

Johnson, N. R. and W. E. Feinberg. 1977. "A Computer Simulation of the Emergence of Consensus in Crowds." *American Sociological Review* 42:505–21.

———. 1989. "Crowd Structure and Process: Theoretical Framework and Computer Simulation Model." In *Advances in Group Processes*, Vol. 6, edited by E. Lawler. Greenwich, CT: JAI Press.

Kosaka, K. and T. J. Fararo. 1991. "Self-location in a Class System: A Formal-Theoretical Analysis." Pp. 29–66 in *Advances in Group Processes*, Vol. 8, edited by E. Lawler. Greenwich, CT: JAI Press.

Lave, C. A. and J. G. March. 1975. *An Introduction to Models in the Social Sciences.* New York: Harper.

Leik, R. K., R. E. Anderson, and G. A. Gifford. 1987. *The Social Power Game.* New York: Random House.

Markovsky, B. 1987. "Toward Multilevel Sociological Theories: Simulations of Actor and Network Effects." *Sociological Theory* 5:100–15.

Markovksy, B., D. Willer, and T. Patton, 1988. "Power Relations in Exchange Networks." *American Sociological Review* 53:220–236.

Marsden, P. V. 1983. "Restricted Access in Networks and Models of Power." *American Journal of Sociology* 88:686–717.

McPhail, C. 1991. *The Myth of the Madding Crowd.* New York: DeGruyter.

Skvoretz, J. and T. J. Fararo. 1992a. "Power and Network Exchange: An Essay toward Theoretical Unification." In *The Location of Power in Exchange Networks* (a special issue of *Social Networks*), edited by D. Willer.

Skvoretz, J. and D. Willer. 1991. "Power in Exchange Networks: Setting and Structural Variations." *Social Psychology Quarterly* 54:224–38.

Willer, D., ed. 1992. *The Location of Power in Exchange Networks* (Special Issue of *Social Networks*).

Willer, D., B. Markovsky, and T. Patton. 1989. "Power Structures: Derivations and Applications of Elementary Theory." In *Sociological Theories in Progress: New Formulations*, edited by J. Berger, M. Zelditch, Jr., and B. Anderson. Newbury Park, CA: Sage.

Wilson, E. O. 1975. *Sociobiology.* Cambridge, MA: Harvard University Press.

Yamagishi, T., M. R. Gillmore, and K. S. Cook. 1988. "Network Connections and the Distribution of Power in Exchange Networks." *American Journal of Sociology* 93:833–51.

ACTION, SOCIAL RESOURCES, AND THE EMERGENCE OF SOCIAL STRUCTURE:

A RATIONAL CHOICE THEORY

Nan Lin

ABSTRACT

The paper outlines a rational choice theory, with the focus on the problematic—how rational actions lead to social structure. My basic arguments are three-fold. First, rational action is seen as having multidimensional motives regarding valued resources. At least two are considered fundamental: minimization of loss and maximization of gain. These are independent, though empirically correlated, calculations, with the former claiming priority over the latter. Secondly, these calculations, and the problem of succession, lead to rules of resource transfers and the primacy of the primordial group. Interactions and collective action in the primordial group are guided primarily by the sentiment to retain and defend resources and secondarily by the need to gain. Third, in general, the utility of social resources, resources embedded in social ties, substantially exceeds that of personal resources. This calculation, in face of

Advances in Group Processes, Volume 11, pages 67–85.
ISBN: 1-55938-857-9

scarcity of valued resources, propels the extension of interactions beyond one's primordial group. Once such ties and exchanges are formed, certain collective rules follow. These rules, beyond the original intent and interest of interacting actors, constitute the basis for the formation of a social structure.

SOCIOLOGICAL THEORIZING

One way to categorize theorizing in sociology is to capture how a theory specifies its causing and consequent concepts relative to two levels of society: structure and actors. If these two levels constitute a dichotomy, a simple typology may look like that presented in Table 1. This typology identifies four types of theory. A macro-theory specifies both causal and effectual concepts at the structural level. A micro-theory posits a relationship between causing and effectual concepts at the actor level. A structural theory is one which links causing structural concepts to effectual actor-level concepts. And an action theory hypothesizes structural effects of actor-level concepts.

This is a simplification, because it is possible to specify a more complex theory that involves: (1) causal or effectual concepts at both structural and actor levels; or (2) concepts implicating more than two levels (e.g., individual actors, organizations, and society; see Hannan 1992). For example, a theory concerning an actor's psychological well-being (an actor-level effectual concept) may be specified as a consequence of both her or his network support (a structural-level concept, and his or her self-esteem (an actor-level concept) (see, for example, Lin, Dean, and Ensel 1986). Likewise, a theory may concern the level of income as a consequence of the level of education (an actor-level concept), the nature of the firm (an organization-level concept), and the industrial sector (an economy or society-level concept) (see, for example, Kalleberg and Lincoln 1988).

Given these precautions, the typology in Table 1 informs us the fundamental theoretical stance within which a particular theory positions itself. My sense is that of the four types of theorizing, action theory is the most challenging and controversial. It is challenging because its causal concepts clearly intersect those primarily and usually identified as under the domains of other scientific disciplines: econom-

Table 1. A Typology of Sociological Theorization: Based on Macro-Micro Specification

	Causal Concept	
Effectual Concept	*Structure*	*Actors*
Structure	Macro-theory	Action Theory
Actors	Structural theory	Micro-theory

ics, psychology, or cultural anthropology. The rational choice theory, for example, extensively borrows the economic assumptions concerning optimization or maximization of choices relative to self-interest (Coleman 1990). Psychological and personality characteristics lay claim to concepts such as well-being, distress, and attitudes (see discussion about shame, Elias 1978; and emotion, Scheff 1992). Norms, values, and traditions can hardly be disassociated from collective and socialization experiences (Marini 1992). An action theory does not wish to disown these potential sources of action (or spring of action, as Coleman calls it 1990). It merely considers them as factors exogenous to the theory. The theory nevertheless needs to demonstrate that it involves more than a simple derivation from concepts already claimed theoretically by other disciplines.

Action theory is also controversial, because its principal proposition concerning the causal linkage from action (the actor-level concept) to structure seems to suggest that the whole can be explained by interacting parts. In general, trans-level causation is harder to demonstrate theoretically than same-level causation. The structural theory, however, has at least the advantage of the omnipresence of structure over actors. Thus, when it is claimed that an actor's job seeking behavior is dictated by the tightness of the labor market, it is hardly possible to place such actions outside the context of the labor market.[1] Action theory, on the other hand, does not have the advantage because it is generally assumed that the structure is more than the sum of actions and interactions of actions (see the argument of structural or organizational robustness, Hannan 1992). Further, once a structure is in place, it becomes theoretically difficult to rule out the continuous interaction between structure and action. An action theory faces the constant challenge to demonstrate whether and how effects of action remain when or after structural effects are taken into account. Abell (1992) correctly points out that the primary puzzle for a rational choice theory is to demonstrate how "interdependent individual actions produce system (or collective) level outcomes."

The present paper sketches a theory proposing how actions may lead to the emergence of social structure. I choose to theorize this process because it should theoretically (logically) precede processes dealing with interdependence and mutual causation between structure and action. Once the issue of action leading to the emergence of structure is "solved," then interdependence and interaction between the two should follow (action affects structure and structure affects action). By focusing on the issue of an emerging social structure, I hope to shed light on other critical issues involved in action theory: what rationality is, what principles guide action and interaction, and why social structures (group and collectivity) are not only possible but inevitable from such action and interaction principles.

My basic arguments are three-fold. First, rational action is seen as having multidimensional motives regarding valued resources. At least two are considered fundamental: minimization of loss and maximization of gain. These are independent, though empirically correlated, calculations, with the former claiming priority over the latter. Secondly, these calculations, and the problem of succession, lead to rules of resource transfers and the primacy of the primordial group. Interactions and collective action in the primordial group are guided primarily by sentiment for retention and defense of resources and secondarily by need for gain. Third, in general, the utility of social resources, resources embedded in social ties, substantially exceeds that of personal resources. This calculation, in face of scarcity of valued resources, propels the extension of interactions beyond one's primordial group. Once such ties and exchanges are formed, certain collective rules emerge. These rules, beyond the original intent of interacting actors, constitute the foundation for the formation of a social structure.

The paper first introduces the basics of the theory of social resources, and proceeds to offer some fundamental propositions concerning action and interaction. These basic propositions, in conjunction with assumptions about succession, lead to conclusions regarding the formation and significance of the primordial group. The paper will then specify the relative utilities of personal and social resources and argue that the relative utility of social resources constitutes a motive for interaction and exchange with actors outside the primordial group. The paper concludes with some further discussion on the nature of the emerging social structure.

SOCIAL RESOURCES THEORY: IMPLICATIONS FOR ACTION AND INTERACTION

In previous work (Lin 1982), I have proposed that actions are driven for the purpose of either preserving (maintaining) or gaining resources. Resources are defined as material and symbolic goods. For the purpose of the present discussion, the focus will be on valued resources. Valued resources are resources individually and, subsequently, consensually deemed useful for survival. In generalized terms, valued resources refer to goods associated with wealth, status, and power (Weber 1957; Lin 1982).

Resources are of two kinds: personal and social. Personal resources are in the possession of the actor who commands the full property rights to them (for discussion about the nature and specification of property rights and private property, see Alchian and Demsetz 1973; Willer 1985). Social resources are resources embedded in one's social network (e.g., wealth, status, and power of social ties).

When action is taken to maintain or defend resources, it is called expressive action, because such action represents a means with no separate end (gaining resources). When action is taken to gain additional resources, it is called instrumental action, for it refers to a class of actions taken as means to achieve goals distinguishable from the actions themselves (additional resources).

For instrumental actions, access to and use of social resources are more critical than the use of personal resources. This was defined as *the social resources proposition*. The reasoning for this proposition is that for instrumental action, intending to gain resources that are not already in possession by the actor, an appropriate strategy would be to access a social tie who occupies a higher position in the hierarchical social structure. It is assumed that higher social positions not only command more resources but also hold a better vision and command of the positions in the structure, especially those below one's own. Thus, the higher the social tie evoked, the greater the tie's resources and command of other positions (social resources) become and the more likely they would be useful to help achieve ego's action goal, resulting in his or her gain of resources.

Other propositions specified causal factors leading to better access of social resources: original social position (the strength of position proposition) and extensiveness of social ties (the strength of ties proposition). Details of the social resources theory and supporting evidence for the propositions can be found elsewhere (Lin 1982; Lin, Ensel, and Vaughn 1981; Marsden and Hulbert 1988; de Graaf and Flap 1988).

What is of interest here is how such actions can be analyzed relative to social interactions (Lin 1990). A major proposition in sociology explaining patterns of social interactions is the like-me or homophily principle which states: social interactions tend to take place among actors sharing resources and characteristics (Lazarsfeld and Merton 1954). Frequency of interaction is seen as constrained within a structural range. Thus, according to this theory, "normative" (in terms of likelihood or relative frequency of) interaction tends to take place among actors with similar statuses and lifestyles. A further extension of the theory links homophilous interaction, to sentiment or attitude (Homans 1950): Homophily of interaction promotes sentiment, and sentiment promotes homophily (in the lifestyle sense) of interaction.

A joint consideration of the types of action and the patterns of interaction, as specified previously, suggests then that the normative (homophilous) interaction is consistent with expressive actions. Actions taken to maintain resources and interactions with others who hold similar resources should be positively related. If this deduction is logically correct, then instrumental action requires "extraordinary" interactions. That is, obtaining additional or better resources requires interacting, directly or indirectly, with actors that have resources or access to such resources, implying actors in different social positions than ego's, thus the heterophilous interaction. In derivation, we may argue that instrumental actions are relatively infrequent and require reaching out to actors outside of a normative social circle (see Granovetter 1973, 1974, 1982 regarding the strength of weak ties argument).

These considerations and deductions inform us of at least two things. For one, it suggests that preferences in actions are dictated by certain needs—either expressive or instrumental. That is, self-interest is at least two-dimensional. The interactions among actors, or interdependence of actors, must take into account both dimensions. A theory, such as the traditional rational choice theory, would concern only seeking resources as the motive and assume the economic calculation as the primary basis for interactions. Another theory may concern maintaining resources as the prime motive, rendering its focus to interactions promoting sentiment and sharing. We could argue that this is also a rational choice theory. But the rationality is now defined beyond the economic principle, and certainly does not follow the maximization principle. Here the concern is the defense of one's resources, a defense against loss, rather than as an "offense" to seek gains. Much research evidence (Marini 1992) supports the claim that in routine behavior, defense against loss may be a greater concern than seeking of gain.

In sum, theoretical developments and research evidence support the views: (1) that defense of resources is a legitimate and top-priority motive for action; (2) that such actions tend to foster interactions among homophilous actors; and (3) that such interactions promote sentiment and sharing of resources. In other words, expressive action takes precedent over instrumental action.

In any event, what emerges from these discussions is that rational choice theory may be valid, so long as rationality is conceived of as having multidimensional drives or springs. The drive to defend against loss must have a central place in the sociological approach to rational choice theory, unlike the economic approach.

So far, the social resources theory has been developed and research undertaken to understand meaningfulness of actions within the context of social structure. That is, the theory has addressed the issue of actions, while acknowledging and recognizing the a priori existence and effect of social structure. What I propose to explore in this paper is to pursue the plausibility that actions may lead to social structure. That is, I seek to develop some theoretical arguments to answer the question: whether rationality based on maintenance or defense of resources as well as expansion and gaining of resources allows us to better understand rules of interaction, and the formation of primary social groups (e.g., the primordial group). And further, whether consideration of the relative utility of social resources to personal resources offers the theoretical plausibility that rational actions may indeed lead to the emergence of social structure, beyond the primordial group.

These explorations are speculative in nature, and will be inevitably brief here. The purpose, nevertheless, is to present the key arguments and to outline a set of propositions so that further elaboration and evaluation is possible.

PRINCIPLES OF ACTION:
MINIMIZATION OF LOSS AND MAXIMIZATION OF GAIN

The theory begins with two simple assumptions about motives for action: (1) that actions, primarily and foremost, are driven or motivated by the innate need for survival; and (2) that survival is seen as dependent upon the accumulation of valued resources. These assumptions regarding the motives (springs) for action require no further elaboration. What needs to be explored are the principles for action— choices and priority among choices. Again, for simplicity, I assume that action is driven either to defend (maintain) resources or to seek (expand) resources.[2] Action driven by defense of resources is a calculation for minimizing loss of resources (relative loss to cost). Action driven by expansion of resources, on the other hand, is a calculation for maximizing gain of resources (relative gain to cost). Based on leads from previous sociological theories (the homophily principles and predominance of expressive needs) and research evidence regarding the relative significance of assessing losses and gains, I now propose that defending resources has higher priority than expanding resources:[3]

Proposition 1. Defense and maintenance of resources are the ultimate motive for action. Thus, *the first principle of action is a calculation of minimizing (resource) loss.*

Proposition 2. Gaining and expansion of resources are the next primary motive for action. Thus, *the second principle of action is a calculation of maximizing (resource) gain.*

These propositions present two important arguments. First, minimization of loss and maximization of gain are two different functions rather than an inverse function of each other.[4] They may involve different choices (what kind and how much of a resource) and, therefore, different preferences. Second, they form a ranked action set, rather than dichotomy. A series of actions may manifest multiple motives: to minimize loss and maximize gain. Given the opportunity, actions are taken to fulfill both motives. However, when confronted with a situation in which the actor must make a choice, the preference is given to maintaining resources: the higher priority is given to the calculation that minimizes loss.

RECOGNITION AND PROFIT: PRINCIPLES OF INTERACTION

How would these two action principles implicate interactions? They would, first of all, suggest that interactions are foremost engaged for the purpose of minimizing loss of resources and secondarily maximizing gain. An interaction following the principle of loss minimization strives to defend loss of resources to another actor. The best possible outcome is that there is no loss. If both actors employ the minimization principle, one local equilibrium is that both actors accept the no loss outcome for both actors. In social terminology, this outcome is a *mutual recognition* of each other's claim to their respective resources—property rights.[5] Recognition, therefore, is a cost to each actor in that ego abandons any challenge to the alter's sovereignty over its resources.[6] It is a minimal cost.

This is a local equilibrium, because it is very constrained. First, it assumes that only two actors engage in the interaction. When multiple actors (three or more) are engaged, coalition is likely to result and the local equilibrium becomes increasingly difficult to maintain. Second, it is seldom the case that the two actors bring equal resources to bear in the interaction. Thus, recognition itself becomes a variable rather than a constant. That is, recognition may come about with unequal costs to the two parties. One actor may be willing to give more recognition to the alter in that ego not only disclaims the alter's sovereignty to resources, but also commits itself to come to the defense of the alter, should its sovereignty be challenged in interactions with other actors. Or, recognition may be maintained only after an actor has also agreed to give up some resources to the alter. Thus, at minimum, there are two types of recognition. In the first instance, where mutual recognition is achieved with minimal cost to each actor, we may consider the recognition as approval or social approval (Lindenberg 1992). In the second instance, recognition involves legitimation—certain generally accepted rules for responsive actions to assure recognition. Third, seldom do actors use the pure minimization principle in a series of actions. Recognition may be a temporary outcome—until one or more actors proceed to evoke the principle of maximization of gain.

Thus, in realistic situations, recognition usually comes as an outcome with unequal costs to parties. Nevertheless, I argue it is the fundamental principle for interactions, for it guarantees the minimal survival of an actor and is consistent with the first principle of action (Proposition 1).

Proposition 3. *Interaction, following the minimization principle of action, seeks recognition of one's claim to resources.*

The element of recognition, I argue, is consistent with some concepts acknowledged or developed by several rational choice theorists. For example, Lindenberg's discussion of social approval (1992) can be seen as one instance of recognition. What is made explicit here is that recognition in interactions can be understood when action is motivated by the principle of minimization of loss rather than the principle of maximization of gain.

I will skip further discussion on interactions based on the principle of maximization of gain, for it merely reflects the usual economic calculations as developed extensively in the literature. What needs to be pursued at this point is how these principles of action and interaction offer clues about the emergence of social structure.

SUCCESSION AND TRANSFER OF RESOURCES: THE PRIMACY OF THE PRIMORDIAL GROUP

Human actions are further compounded by additional innate but prominent life circumstances: finality of life and reproductivity of life. Survival of an individual actor is limited in time. One possible consequence of an actor's exit might be that all resources associated with the actor revert back to a pool for other actors to freely compete to gain. However, this strategy would mean a total loss of resources to the actor after life-long efforts (actions and interactions) to maintain and expand them.

Alternatively, the resources may be transferred to another actor(s). An extension of the primacy of the principle of minimization of loss (Proposition 1) suggests that the actor prefers to transfer claimed resources to another actor deemed most suitable as a surrogate. Suitability is reflected in the extent to which the surrogate is easily identified with the actor in continued recognition and legitimation relative to other actors. Reproductivity of life, in most societies, offers an easy rule to identify the surrogate. Thus, for most societies, the primordial group, the family, becomes the immediate and natural extension of the actor.[7]

The primacy of the primordial group for succession and transfer of resources further incorporates noneconomic considerations in actions. Restriction of succession within a primordial group reduces the range of choices of the surrogate. Depending on the rules of succession, the choice may be reduced to zero degree (e.g., the eldest son as the successor). Thus, recognition and legitimation considerations are given increasing priority over competence and skills useful in maximizing gains—the economic calculation. It is clear the existence of the primordial group, as it prevails through human history, makes any theory based on economic calculations alone unattainable.

This last conclusion does not lead to another conclusion: that actions are not rational. If rationality is meant as the process of reasoning by way of calculation over choices, then it is clear, as argued earlier, recognition and profit provide rational bases for interaction choices.

PERSONAL RESOURCES, SOCIAL RESOURCES AND SOCIAL NETWORK

The need to minimize loss and to maximize gain establishes two building blocks to understand interactions beyond the primordial group. However, we need to introduce another building block: consideration of the relative utility of two kinds of resources: personal and social.

Personal resources are resources in the possession of the actor who can exercise the decision (authority) in their usage and disposition. These possessed goods can also be transferred to designated successors as the actor sees fit. *Social resources* are resources attached to other actors. Interactions and relations with other actors offer the possibility that such resources can be "borrowed" for ego's purposes. In return, the borrowed resources must be returned, replaced, or reciprocated. In the most primitive terms, borrowing a neighbor's cutting instruments during harvest would be one such example of the access to and use of social resources. Once the action, harvesting, is accomplished, the instruments are returned (either intact or replaced) to the neighbor and, more importantly, the expectation is that the neighbor may borrow ego's resources, such as his son, to help him harvest as well.

Because of the constraints attached to the use of social resources and the energy and resources required in the maintenance of the relations and reciprocal transactions, sentiment dictates preference for the accumulation of personal resources to that of social resources. That is, the relative cost (temporality in use, obligation for return or replacement, and commitment for reciprocity) for using personal resources is much lower than that for using social resources. Then, how do we account for the use of social resources and, therefore, maintenance of social relations? That, of course, is the critical and pivotal issue in any theory linking actions to structure.

The pinnacle argument, for the author, rests with two central theoretical propositions:

Proposition 4. *The accumulation of social resources is much faster than that of personal resources.* That is, *accumulation of personal resources tends to be additive* in nature, *whereas accumulation of social resources tends to be exponential* in nature.

Proposition 5. *When interactions outside one's primordial group are intended for gaining resources, they are more for accessing social resources than gaining personal resources.*

Personal resources are accumulated by actions taken by the actor and members of his or her primordial group. Each action generates a given amount of additional resources. Therefore, there is a tendency to expend the primordial group (e.g., the extended family) so that the generation and accumulation of resources can accelerate.

Social resources, on the other hand, are generated by making and maintaining of social ties. A relation with a social tie suggests a linkage and, therefore, access to the tie's resources, social resources for ego. Further, once a tie is accessed, not only do his or her resources become social resources to ego, but also social ties of the alter offer possible social resources. Conceivably, social resources might be accessible through ego's network of direct and indirect ties. The extent of access to such social resources, of course, depends on how much resources are at the disposal of the social ties as well as on the nature and extent of the ties. As these ties extend into a network of both direct and indirect ties, the pool of social resources accelerates exponentially. Thus, by the sheer networking principle, the potential pool of social resources becomes extended quickly. The hypothesized rates are depicted in Figure 1.

Further considerations need be given about possible models of exponential accumulation of social resources. The free rendition in Figure 1 is entirely conjectural. The slope of the S-shaped curve is based on the assumption that interactions and networking extend first slowly, probably among a small number of actors of similar resources, and then quickly, to larger numbers of actors with dissimilar and better resources as the network extends through indirect ties. It plateaus and reaches an upper limit, because the function must be constrained by an efficiency factor (it may be a function of number of intermediary links, associated negatively with recognition and legitimation and positively with cost or multiplicity of reciprocal

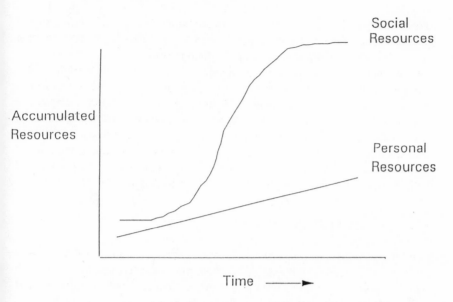

Figure 1. Accumulation rates for personal and social resources

obligations). While the relative cost of accumulating and using social resources is high, there are conditions when such cost is more than compensated for and exceeded by the relative advantage in the speed of accumulating social resources. The calculation tipping in favor of social resources approaches inevitability for most actors due to the likelihood of two limiting conditions for the accumulation of personal resources: the ultimate limiting size of the primordial group and scarcity of materials for resources.

As mentioned above, one strategy of speeding up the accumulation of personal resources is to expand the primordial group, whose membership shares the interest in the production and transfer of resources. However, as the size of the primordial group increases, it also creates problems for the maintenance of centralized authority over resources and competition for the succession to the entitlement of the resources. To maintain the expansion of the primordial group, more and more resources must be generated so that there is sufficient sharing among the members. So long as there are plenty of raw materials for the production of resources, the expansion of the primordial group can continue.

However, it is inevitable that multiple groups seeking resources increase in size and numbers to a point where they will have to compete for materials. Competition for scarce materials can and does get primitively resolved by one group physically taking possession of other primordial groups, turning the members of the latter into resource-generating instruments—enslaved laborers. However, unless the ability to take possession of another group is overwhelming in terms of relative size or superiority of instruments (technology), there is always a risk that the confrontation will result in the enslavement of the ego's group instead.

An alternative to the competitive resolution in face of scarcity of materials is the access to and use of social resources, even though such use, as discussed, incurs a greater cost than the use of personal resources. Once such a rational decision is made, it is not only the case that interactions with actors beyond one's primordial group take place but are actively sought after, in order to access social resources. Such access is also entirely consistent with the motives for actions—minimizing loss and maximizing gain, and the principles of interactions—recognition and profit. Social resources can be mobilized to accomplish both purposes.

THE EMERGENCE OF STRUCTURE

Such access comes at an important cost—willingness and preparedness to reciprocate in terms of recognition and profit. There must be a commitment to provide one's own resources as social resources to others. To assure stable access to social resources and to demonstrate reciprocity, interactions are routinized—that is, the formation of social relations. The maintenance of social relations, likewise, is based on the two rational principles as specified in Propositions 1 and 2. Social resources are foremost used and relied on to maintain, sustain, and defend one's resources. They are, secondarily, used to gain additional resources. Legitimation guides reciprocity and the calculation process. The calculation is complicated by the fact that routinized social relations directly and indirectly involve multiple actors and their primordial (and extended) groups. While such relations promote access to social resources, recognition and legitimation of the relations and obligations elevate quickly in complexity for calculations. That is, sharing social resources and an increasing need for legitimation rules go hand in hand. In subsequent actions, an actor's calculation must take into account whether the action is consistent with the obligation to defend and/or expand the resources of the interacting actors.

The multiplicity and complexity of routinized social relations demand increasing rules of recognition and legitimation. These rules recognize the basic rights to personal resources (property), and, at the same time, specify responsibilities and obligations for actors to contribute resources.

COLLECTIVITY AND ITS RULES

Once such social relations and sharing of resources are established and maintained, there is a collectivity. A collectivity is an aggregation of actors and primordial groups bound together for the sharing of social resources. A collectivity can also decide to produce further resources that belong to the collectivity rather than any specific actors, the public resources. The persistence of a collectivity depends on a set of formal and informal rules governing actors relative to each other and to the access and use of shared resources whether they are social or public. These rules establish differential obligations and rewards for member actors.

Differential obligations are necessary because the persistence of the collectivity depends on the maintenance and gaining of shared resources. Obligations include two types: (1) recognition and loyalty (sentiment) to the collectivity and its rules and (2) amount and type of performance (work) in the production of shared resources, especially the public resources. The loyalty factor minimizes the loss of collective resources. The performance requirement maximizes the gain of such resources.

Differential rewards are necessary because actors are evaluated as differentially fulfilling their obligations to the collectivity. Thus, more rewards are given to those who demonstrate a high degree of loyalty and/or high level of performance. Rewards can be symbolic as well as materialistic. Materialistic rewards include the designation and allocation of resources to the actor (the gaining of personal resources) and the access and use of shared resources (social and public resources). Symbolic rewards include public commendation of the actor and assurance of transfer of such honors to future generations of the actor. Another increasingly important reward system concerns rules and procedures for allocating enforcing-agent positions in the collectivity. This will be further discussed.

These obligations and rewards, while required for the persistence of a collectivity, both complement and compete with the primitive obligations actors have for themselves and their primordial groups. They are complementary because the shared resources in the collectivity supplement personal resources, so that shortage of personal resources no longer need always be survival threatening. They are competitive because allocation of energy for the production of resources and commitment of loyalty can be taxing for meeting the dual needs.

There is the inevitable conflict of interest, however. Since the primitive motives drive the actors to the maintenance and gaining of personal resources rather than social resources, willingness to perform and to be loyal to the collectivity and collective goods depends on at least two important factors: (1) how important the social resources are to the actors; and (2) how collective obligations and rewards, in terms of loyalty and performance, synchronize with primary obligations and rewards. The more positive the two evaluations, the more likely the actors will be willing to perform and be loyal to the collectivity and the collective goods. In the extreme situation, an actor may be willing to make the ultimate sacrifice, his or her own life, in order to preserve shared resources for his or her primordial group and the collectivity.

If the two factors are not seen as matched, two outcomes are likely. One may choose to exit from the collectivity, at the risk of losing social resources but in the hope of finding another collectivity better matched for the interest of ego and his or her primordial group. Or, there is an increasing likelihood of becoming a free rider, who takes and treats shared resources as personal resources.

There are of course risks associated with these choices. Exit increases the problem of protecting and finding resources for survival. Free riders may run the risk of punishment (deprivation of personal and social resources) established by the collectivity, which will be discussed later.

As the size of the collectivity increases, interactions become fragmented (localization of networks) and shared resources become segmented (localization based on shared resources and characteristics). At the collective level, obligations and rewards must be continuously revised to cover the increasing number of actors and their needs for social resources. As a result, the proportion of collective obligations and responsibility overlapping with individual actors and their primordial groups decreases. Routinized recognition and legitimation will decrease in their utility to bind actors to the collectivity.

SOCIAL CONTRACTS

To assure that collective obligations and rewards are perceived as matched with those of member actors' own, that structural problems of fragmentation and segmentation are overcome, that loyalty and performance are exercised and that exits and free riding minimized, a collectivity can develop and employ three strategies: (1) cultivate through education and culture to *internalize* collective obligations and rewards (Marini 1992); (2) engage in mass campaigns promoting the *identification* of the actors with the attractiveness of shared resources and the collectivity; and (3) develop and enforce rules for *forced compliance*. Kelman's discussion on the three processes (internalization, identification, and compliance) (1961) implies that these three strategies form points along two axes. Compliance can be achieved with maximal speed, but minimal span of effect. When control is present, compliance is quickly achieved (e.g., war prisoner's behaviors). But when it is absent or lifted, such behaviors will also quickly change or disappear.

Internalization, on the other hand, takes the longest time to achieve, but the consequent behaviors presumably persists with minimal control. Discussion of the employment of these strategies is beyond the scope of this paper. What needs to be emphasized is that each of these strategies entails the development of rules of engagement of actors in the collectivity. Further, agents and agencies of enforcement must be developed.

These enforcing actors are to administer and manage the activities as well as enforce the rules of the collectivity. The emergence and necessity of these enforcing agents generate further relationships between actors and the collectivity. These agents assume authority over individual resources and act on behalf of the collectivity. While they are expected to defend and expand individual actors' resources as well, their ultimate rewards come from demonstrations of their loyalty to the collectivity and expansion of the public resources.

As scarcity of resources increases and the collectivity increases in size, the enforcing agents gain prominence among members, as the survival of the collectivity increasingly depends on the agents' enforcement of rules. Thus, a hierarchy among actors will emerge not only because of differential obligations and rewards but also because of differential allocation and opportunity to be enforcing agents.

SOURCES OF TENSION IN SOCIAL SYSTEMS

Space does not permit further elaboration of relationships between corporal (enforcing) actors and natural actors, the formalization of legitimation and profit rules in the realized social system, and the perpetual tension between loyalty and profit for a social system and its individual and corporal actors. However, we can point out several sources of tension in social systems that ensue. The most obvious one is the tension between personal resources and social resources. Because of the ultimate survival instinct, and the cost of accessing social resources, there is a much stronger tendency for a natural person to strive for personal resources. A social system needs to strike a balance between allowing opportunities for the participants to maintain and gain reasonable levels and amounts of personal resources and enforcing their willingness to produce and maintain social and public resources.

A second source of tension is the balance of mobility versus solidarity. Mobility represents the opportunity to move upward in the social hierarchies whereas solidarity is the need to identify with other participants with similar interests and resources.[8] Mobility encourages breaking away from one's social circle of shared interests and resources in order to gain more or better resources in the social system. Solidarity relies on identification with others sharing similar resources and sentiments. Overemphasis on mobility tends to break down social identity and group cohesion. Overemphasis on solidarity fragments sectors of the structure and creates potential class identification and conflict. How to strike a balance between the two is critical for the survival of the social system.

Still another source of tension originates from size of the system. One consequence of the increasing size is decreasing shared resources relative to the amount of resources "unique" to member actors. Thus, the value attached to the commonality of shared resources decreases for the members. This creates a tendency for member actors to form subsets of relations with others who share resources of common interest and value. Special interests and lobbying efforts by the subset of actors and collectivities competing for rules in their favor can tip the legitimacy of the rules regarding distribution of shared, especially public, resources available to the system. As shared resources become relatively more scarce, these competitions, if unchecked or not resolved, could lead to fragmentation of loyalty. Loyalty would be shifted to groups or clusters within the system rather than the structure as a whole, endangering the identity and persistence of the system as a whole. How to maintain the structure while it continues to grow in size and facing increasingly shared resources is an issue no open social system can avoid (see similar discuss in Coleman 1986a, 1986b).

CONCLUDING REMARKS

In this paper, I propose that rational principles for action be conceived of as two types, minimization of resource loss and maximization of gain, with the former claiming primacy. This position does not challenge the primacy of action, nor the viability of rationality as a theoretical argument. It does challenge the exclusive use of the economic profit-maximization (or even profit-optimization) approach as the sole basis of accounting for human actions, interactions, and the functioning of social organizations. Further noneconomic but quite rational calculations naturally and logically flow from issues fundamentally linked to the nature of human life, such as reproduction and succession, sovereignty of property, and the need for recognition of such sovereignty—issues that any theory about human society cannot ignore, yet the economic approach does.

Consideration of these issues does not relegate sociology's significance to psychology or cultural anthropology. Claims to property rights, recognition, transfer of resources, and succession clearly are all socially driven. They describe social life and social activities and are only meaningful in interactive and networking contexts.

Not only does rationality for action spring from the innate nature of human life, but also principles of interactions cannot afford to ignore two different types of resources: personal and social. A model that considers exclusively transactions of personal resources will never be able to account for the links between actors and social structure, because social networks and social resources are at the core of the micro-macro link. Concepts of power, dependence, solidarity, social contracts, and multi-level systems would not make sense until social resources are brought into consideration.

The present paper demonstrates how several simplified propositions concerning principles of action and interaction thus conceived can explain the emergence of social structure from bases of action and interaction: an action theory of society. The propositions and theoretical arguments presented here, I believe, provide building blocks for further analysis of formation and development of social institutions and organizations. For example, considerations of social contracts can extend to multiple social contracts and the subsequent hierarchical structure subsuming these contracts by way of variations in the social (recognition), political (legitimation), and economic (profit) rules.

Once a social system is in place, it inevitably becomes the dominant aspect of social life. Its imposition on individuals is increasingly pervasive. Therefore, we must necessarily take structural effects as given when we describe observable social systems. I agree with Hannan's observation (1992) that organizations take on characteristics unintended and unpredictable from individual actions. However, the principle for the "robustness" of social systems, I believe, is derivable from the same principles guiding individual actions and interactions. That is, principles of minimization of loss, maximization of gain, rules of resource transfer and succession, and primacy of social (public and shared) resources over personal resources, guide institutions and organizations to establish rules in their authority structure, opportunity structure, and sociocultural structure as well. Collective interest will supersede individual interest, just as primordial group interest supersedes individual actor interest. Loyalty will supersede performance in reward/punishment rules as recognition supersedes profit for individual actors. While the principles are similar, the primacy of collectivity over individuals forges "structural" variations not accountable from individual action and interactions.

Ultimately, a viable social theory must integrate both individual and structural elements. A comprehensive and balanced treatment of these two elements, I suspect, is the challenge sociologists must accept in order to offer theories that are both analytically and descriptively valid.

ACKNOWLEDGMENT

An earlier version was presented at the XII World Congress of Sociology, Madrid, Spain, July 1990. Support from the National Science Foundation, the Trent Foundation, and Duke University Research Council for sustaining this work is gratefully acknowledged.

NOTES

1. This is not to argue or claim for the validity of the proposition.
2. This assumption is consistent with the purposive action approach.
3. This is valid only if the actor has some resources to begin with.
4. Empirically, they may be negatively correlated.
5. Trust may be an alternative term. However, I believe recognition may be evoked without trust which has a stronger affective meaning.
6. Note that this would have been considered as a mere stalemate or worst outcome if the principle of maximization is evoked.
7. Elsewhere (Lin 1989), I discuss rules of transfer and types of resources transferred. There are variations in transfer rules within the context of family. For example, inheritance rules vary across societies and there is no uniformity regarding unigeniture, primogeniture or even-distribution principles, even though there seems a strong tendency toward male primacy. In the most interest case, the Chinese traditional system uses split rules: primogeniture for authority inheritance, but even-distribution among sons for property inheritance. The resulting conflict and chaos as well as diminishing pooled resources cannot be explained by any economic principles. Nor is the family group, the predominantly primordial group in most systems, the only primordial group. A primordial group can and has been constructed on other criteria (e.g., ethnic, religious, and gender identities). These variations, however, do not affect subsequent arguments on paper.
8. I simply define solidarity as recognition, sentiment and legitimacy expressed by actors in a collectivity who share similar resources and lifestyles. This conception is somewhat different

from Hechter's (1983). Hechter, also embarking from a rational-choice model, suggests that solidarity of a group becomes possible because two elements are present: (1) dependence relations between individuals and the group as determined by access to alternative sources for resources; and (2) a monitoring capacity of the group in terms of both metering individuals' behaviors and sanctioning behaviors via leadership. Thus, Hechter's work can be seen as an attempt to further specify the process of interactions linking individuals to obligations and reciprocity and thereby to various market, authority and norms systems which Coleman (1985, 1986) suggested but never specified. The first element identified by Hechter is a direct application of the dependence-power theory advanced by Emerson and Cook (Emerson 1962; Cook, Emerson, Gillmore, and Yamagishi 1983, 1988). It emphasizes the significance of networking among individuals and issues of resources deemed valuable to the individuals. It can follow from the basic argument of individuals seeking maximal resources through interactions with multiple actors. The second element, the monitoring capacity of the group via leadership, however, creates a component that cannot be derived from and sustained by a rational-choice model. The emergency of leadership is left unexplained and the group monitoring capacity, without stipulation of its potential origin from interaction processes, simply reestablishing the structural constraint and explanation model.

REFERENCES

Abell, P. 1992. "Is Rational Choice Theory a Rational Choice of Theory?" Pp.183–206 in *Rational Choice Theory: Advocacy and Critique*, edited by J. S. Coleman and T. J. Fararo. Beverly Hills, CA: Sage.

Alchian, A. and H. Demsetz. 1973. "The Property Right Paradigm." *Journal of Economic History* 33:16–27.

Coleman, J. S. 1986a. "Social Theory, Social Research: A Theory of Action." *American Journal of Sociology* 91:1309–35.

———. 1986b. *Individual Interests and Collective Action*. Cambridge, MA: Cambridge University Press.

———. 1990. *Foundations of Social Theory*. Cambridge, MA: Harvard University Press.

Cook, K. S., R. M. Emerson, M. R. Gillmore, and T. Yamagishi. 1983. "The Distribution of Power in Exchange Networks: Theory and Experimental Results." *American Journal of Sociology* 89:275–305.

———. 1988. *The Structure of Social Exchange*. Orlando, FL: Academic Press.

de Graaf, N. D. and H. D. Flap. 1988. "With a Little Help from My Friend." *Social Forces* 67:453–472.

Elias, N. (1939) 1978. *History of Manners*. New York: Pantheon.

Emerson, R. M. 1962. "Power-dependence Relations." *American Sociological Review* 27:31–40.

Granovetter, M. 1973. "The Strength of Weak Ties." *American Journal of Sociology* 78:1360–80.

———. 1974. *Getting a Job*. Cambridge, MA: Harvard University Press.

———. 1982. "The Strength of Weak Ties: A Network Theory Revisited." Pp. 105–130 in *Social Structure and Network Analysis*, edited by P. V. Marsden and N. Lin. Beverly Hills, CA: Sage.

Hannan, M. T. 1992. "Rationality and Robustness in Multilevel Systems." Pp. 120–36 in *Rational Choice Theory: Advocacy and Critique*, edited by J. S. Coleman and T. J. Fararo. Newbury Park, CA: Sage.

Hechter, M. 1983. "A Theory of Group Solidarity." Pp. 16–57 in *The Microfoundations of Macrosociology*, edited by M. Hechter. Philadelphia: Temple University Press.

Homans, G. C. 1950. *The Human Group*. New York: Harcourt, Brace.

Kalleberg, A. L. and J. R. Lincoln. 1988. "The Structure of Earnings Inequality in the United States and Japan." *American Journal of Sociology* 94 (Supplement):S121–S153.

Kelman, H. 1961. "Processes of Opinion Change." *Public Opinion Quarterly* 25:57–78.

Lazarsfeld, P. F. and R. K. Merton. (1954) 1982. "Friendship as Social Process: A Substantive and Methodological Analysis." Pp. 298–348 in *The Varied Sociology of Paul F. Lazarsfeld*, edited by P. L. Kendall. New York: Columbia University Press.

Lin, N. 1982. "Social Resources and Instrumental Action." Pp. 131–146 in *Social Structure and Network Analysis*, edited by P. V. Marsden and N. Lin. Beverly Hills, CA: Sage.

———. 1989. "Chinese Family Structure and Chinese Society." *Bulletin of the Institute of Ethnology, Academia Sinica* 65:59–129.

———. 1990. "Social Resources and Social Mobility: A Structural Theory of Status Attainment." Pp. 247–271 in *Social Mobility and Social Structure*, edited by R. Breigher. New York: Cambridge University Press.

Lin, N., A. Dean, and W. M. Ensel. 1986. *Social Support, Life Events, and Depress.* Orlando, FL: Academic Press.

Lin, N., W. M. Ensel, and J. C. Vaughn. 1981. "Social Resources and Strength of Ties: Structural Factors in Occupational Status Attainment." *American Sociological Review* 46:393–405.

Lindenberg, S. 1992. "The Method of Decreasing Abstraction." Pp. 3–20 in *Rational Choice Theory: Advocacy and Critique*, edited by J. S. Coleman and T. J. Fararo. Beverly Hills, CA: Sage.

Marini, M. M. 1992. "The Role of Models of Purposive Action in Sociology." Pp. 21–48 in *Rational Choice Theory: Advocacy and Critique*, edited by J. S. Coleman and T. J. Fararo. Beverly Hills, CA: Sage.

Marsden, P. V. and J. S. Hurlbert. 1988. "Social Resources and Mobility Outcomes: A Replication and Extension." *Social Force* 67:1038–1059.

Scheff, T. J. 1992. "Rationality and Emotion: Homage to Norbert Elias." Pp. 101–119 in *Rational Choice Theory: Advocacy and Critique*, edited by J. S. Coleman and T. J. Fararo. Beverly Hills, CA: Sage.

Weber, M. (1922) 1957. *The Theory of Social and Economic Organization.* New York: Free Press.

Willer, D. 1985. "Property and Social Exchange." Pp. 123–42 in *Advances in Group Processes*, Volume 2, edited by E. J. Lawler. Greenwich, CT: JAI Press.

RELATING POWER TO STATUS

Michael J. Lovaglia

ABSTRACT

Specifying the complex relationship between power and status requires integration of theoretical ideas from three major subfields of social psychology: exchange theories of power, expectation states theories of status, and attribution research. Different formulations beginning with either exchange theories or expectation states theories converge on the proposition that power confers status. However, this is not always the case. Emotional reactions to power use and the effect of attributional biases on expectations of ability can intervene to reduce the effect of power differences on status differences. High power actors gain status from their power to the extent that (a) power use does not arouse negative emotional reactions on the part of low power actors, and (b) information about the structural source of power is lacking, unclear or disregarded. I propose several hypotheses to test these theoretical formulations.

INTRODUCTION

Though questions of power and status are basic to sociology, the relationship between them has evaded specification. There have been preliminary attempts. Kemper and Collins (1990) argued that power and status are the two central,

Advances in Group Processes, Volume 11, pages 87–111.
ISBN: 1-55938-857-9

independent dimensions of microinteraction. Yet it does not seem likely that power and status are totally independent of each other. Powerful people are often high ir status—judges for example—but the correlation is not perfect. A retired ambassador may have high status but little power, and prison guards have power over inmates but get little esteem and honor from them in return. If power and status are not always positively correlated but also are not independent of each other ther how are they related?

Two somewhat contradictory ideas about the relationship between power anc status appear in the classical literature. Weber contended that economic power as represented by property "is not always recognized as a status qualification, but ir the long run it is, and with extraordinary regularity" (Gerth and Mills 1958, p. 187) In the same discussion Weber argued that power does not always confer honor, one aspect of status. "The typical American Boss, as well as the big speculator, deliberately relinquishes social honor" (p. 180). And, Lenski (1966, p. 52) noted, "Honor is deniec to those who rule by force alone." Power confers status but not always.

The relationship can also work in the other direction; status may lead to power and according to Weber frequently does (Gerth and Mills 1958). Ridgeway anc Berger (1986) proposed that if status was legitimated, then high status actors would be able to use power in ways that low status actors would not. Blau (1964) also saw power developing from status differences, as low status actors become increasingly dependent on the contributions of high status actors. While the conditions under which status produces power are also worthy of investigation, I concentrate for the present on ways in which power may confer status.

I use theoretical ideas from several subfields of social psychology to specify the relationship between power and status. Both power and status have large literatures Though other theoretical perspectives may illuminate the relationship between them, common elements among the subfields I chose facilitated integrative work. Exchange theories of power, expectations states theories of status, and attribution research all share a focus on group processes, theoretically driven empirical research, and a largely experimental methodology.

DEFINING POWER AND STATUS

Exchange Theories of Power

Power-dependence theory (Emerson 1962) offers the useful insight that power resides in the relations among actors rather than within actors themselves. Differences in individual capability are small compared to differences in power between positions in a social structure. I use the term, "power," in this narrow, social structural way. In determining the relative power of two actors, Emerson equated the power of actor A with the dependence actor B has on actor A. Thus the main proposition of the theory is that:

$$P_{AB} = D_{BA}$$

where P represents the power of the actor designated by the first subscript over the actor designated by the second. Emerson proposed that inequality of power is inherently unbalanced and tends toward equality through power use.

The theory defines the dependence of B on A as directly proportional to B's motivational investment in goals mediated by A and inversely proportional to the availability of those goals to B outside the $A-B$ relation. A's power stems from the ability to deprive B of a desired goal, either by imposing a cost or withholding a reward. In the extension of power-dependence theory to exchange networks, Emerson (1972) placed additional conditions on what constitutes power and power use. Power in exchange networks requires enduring relations among actors. That is, exchange consists of a series of transactions; each of which is a mutually rewarding instance of exchange. Power is the level of potential cost that A can impose on B. To use power is to impose a cost that unbalances a relationship.

Balance occurs when $P_{AB} = P_{BA}$. Emerson proposed that only balanced relationships are stable. Because unstable relationships are prone to change while stable relationships are not, unbalanced power relations tend toward balance. Emerson (1962) constructed four possible balancing mechanisms when $P_{AB} > P_{BA}$:

1. B reduces motivational investment in goals mediated by A.
2. B cultivates alternative sources of gratification.
3. A increases motivational investment in goals mediated by B (e.g., the stronger actor becomes dependent on the weaker actor as a source of status, a process Emerson called "status giving").
4. A is denied alternative sources for achieving A's goals (e.g., coalition formation).

Emerson (1964) gave experimental support to the main proposition and to balancing mechanisms 3 and 4. Molm (1981) found power imbalance directly related to power use in further support of the theory.

One significant form of power in exchange networks is the availability of alternatives to exchange with a particular partner. Thibaut and Kelly (1959) first emphasized the importance of alternatives for network power. Emerson (1972), in extending power-dependence to networks of actors larger than the dyad, continued this emphasis. While the original theory also relies heavily on behaviorist ideas such as satiation (Emerson 1969), most recent work has focused on "alternatives" as the key to network power (Cook, Emerson, Gillmore, and Yamagishi 1983; Molm 1990). Another influential theoretical program, Markovsky, Willer, and Patton's (1988) Network Exchange Theory rejects the behaviorist aspects of power-dependence entirely. It proposes instead that power in networks is based on "exclusion," a concept closely related to "alternatives." For Markovsky, Willer, and Patton (1988), an actor has power over another to the extent that actor can exclude the other from an exchange relationship. An excluded actor is one who lacks a viable alternative exchange partner.

Definitions of Power in Exchange Networks

I define power here as the social structural capacity to acquire contested rewards. Other exchange formulations use comparable conceptions of power. Markovsky, Willer, and Patton (1988, p. 224) defined power as a "structurally determined potential for obtaining relatively favorable resource levels." I altered their definition in order to keep rewards separate from resources used to obtain them. (Although once obtained, some rewards can be used as resources—money for example.) In exchange research, a "*resource* is any act, attribute, or object that is instrumental to accomplishing the ends of an actor" (Samuel and Zelditch 1989). Power-dependence theorists use a similar definition. For them, power is the capacity to obtain favorable outcomes at another's expense (Emerson 1972; Cook and Emerson 1978; Cook et al. 1983).

Power exists as a capacity but can also be *used* to alter either the fate or behavior of actors. Molm (1990) pointed out that structural power may not directly affect exchange outcomes, such as, the distribution of rewards among network positions. There is no guarantee that an actor will use a power advantage. Further, an actor's attempted power use may be ineffective. Actors strategically use power in ways that alter reward outcomes. Samuel and Zelditch (1989, p. 309) narrowly defined power as power use, "the use of rewards and/or penalties to induce or coerce compliance." Power use in this sense requires that one actor attempt to dominate while another actor submits. For example, an employer asks a worker to submit to a 30 percent pay cut or face lay off during a recession. I will call this *overt* power use to distinguish it from power used to control fate rather than behavior. Overt power use is a special case of power and will have consequences for a theory relating power to status.

Exchange researchers commonly use the amount of rewards obtained in a competitive situation as an indicator of power use and, indirectly, of power itself (Emerson 1962; Cook and Emerson 1978; Cook et al. 1983; Molm 1985; Markovsky et al. 1988).

Expectation States Theories of Status

"Status" refers to an individual's standing in the hierarchy of a group based on criteria such as prestige, honor, and deference. The theories of *status characteristics and expectation states* form a well-established theoretical research program (Berger, Wagner, and Zelditch 1985) that meshes with exchange research on power. Just as Emerson viewed power as relational, the expectation states approach assumes that relationships between actors, not characteristics of actors themselves, produce status differences (Berger, Rosenholtz, and Zelditch 1980). Further, expectation states arise out of social interaction. Because exchange behavior is social interaction, it seems reasonable to proceed with the assumption that exchange behavior can produce expectation states.

The original expectation states theory dealt with the emergence of stable inequalities (power and prestige orders) in problem solving groups composed initially of status equals.[1] It proposes that expectations about the competence of future performances of group members emerge and are maintained during interaction related to a task (Berger 1958; Berger and Conner 1974). Status characteristics theory explains how group members who initially differ in status import these differences into new group contexts with strangers (Berger, Cohen, and Zelditch 1966, 1972; Berger and Fisek 1974; Berger, Fisek, Norman, and Zelditch 1977; Berger and Zelditch 1985).

Status Characteristics Theory

A *status characteristic* is a feature of an actor that influences group members' beliefs about each other.[2] It has differentially evaluated states, that is, one state of the feature is generally considered to be more desirable, highly valued, esteemed, and honored than another. For example, people in many cultures consider light skin more desirable than dark skin. Here skin tone is a cue to the state of the status characteristic, race. Expectation states are *specific* if they apply to performance in a defined situation or *general* if not restricted to a specific situation. Status characteristics are *specific* if they involve two or more states that are differentially evaluated, and associated with each state is a distinct specific expectation state. They are associated with specific performance expectations. For example, chess playing ability is a specific status characteristic. Status characteristics are *diffuse* if (a) they involve two or more states that are differentially evaluated, (b) associated with each state are distinct sets of specific expectation states (each itself evaluated), and (c) associated with each state is a similarly evaluated general expectation state. Gender, occupation, and race are examples. Light skin may be associated with intelligence and thus competence in a wide variety of situations.

The theory limits its scope to task oriented and collectively oriented groups. It proposes that status characteristics determine the status hierarchy of the group, indicated by (a) opportunities given to members to perform, (b) performances of members, (c) communicated evaluations of performances by members, and (d) members' influence on group decisions.

Here I review five logically interrelated assumptions at the core of *status characteristics* theory that connect members' status characteristics to their rank in a group's status hierarchy (Berger, Fisek, and Norman 1989). Actors in the task situation may be interactants or referents. The theory models the interaction process with only two interactants at any one time, a focal actor and an other. A referent is an actor who is not an interactant during a given time period, but whose status information is used by the interacting pair.

By the first assumption, status characteristics become *salient* to group members if they are defined as relevant to the task, or if group members possess different values of the status characteristic. For example, the status characteristic, gender, might be considered relevant in a sex-typed task such as changing the flat tire on a car. However, gender also would be salient when not defined as relevant to the task if the group has both men and women members.

According to the second assumption, group members will act as if salient, status characteristics are relevant unless those characteristics have been specifically dissociated from the task. This has been called *burden of proof* because it implies that actors will consider salient, status characteristics relevant and use them to order interaction unless convinced they do not apply.

The third assumption specifies the *sequence* in which groups restructure interaction when an actor joins a group. The structure develops as the new actor starts a task or engages in interaction with a group member as specified by the salience and burden of proof assumptions.

The fourth assumption models how actors *combine* all relevant status information, though they likely are not aware of the process. All positive status information is combined and given a value, $e+$; all negative status information is combined and given a value $e-$. This combining is subject to attenuation; each additional piece of positive status information adds less to $e+$ than did information that preceded it; each additional piece of negative information subtracts less from $e-$ than did previous information. Aggregated expectations for an actor are calculated by summing $e+$ and $e-$. Subtracting the aggregated expectations of an actor from those of an other gives that actor's aggregated expectation advantage or disadvantage.

The fifth assumption connects expectation advantages and disadvantages to status rank. An actor's position in the status hierarchy of a group relative to an other is a direct function of that actor's expectation advantage or disadvantage relative to that other.

Status characteristics and expectation states theories tie status closely to expectations of ability or competent performance. In task situations where it is expected that all members will contribute to a successful outcome, group members form expectations of each others' competence. These expectations produce a status hierarchy, a finding repeatedly demonstrated in the laboratory and in the field (Berger 1980; Berger Fisek, Norman, and Wagner 1985; Cohen, Lotan, and Catanzarite 1988; Cohen and Roper 1972, 1985; Pugh and Wahrman 1983; Ridgeway 1981, 1984). A member's influence on group decisions is a frequently used indicator of status in these studies.

Distinguishing between Power and Status

Because influence is an indicator of status, it is necessary to distinguish between influence and power to avoid tautology. Festinger (1953) makes a distinction that has proven useful (Samuel and Zelditch 1989; Kemper and Collins 1990). Power can compel public compliance but only the influence resulting from status induces both public compliance and private acceptance. Status, but not power, may cause one actor to defer to another without pressure from bribe or threat. This distinction can be used to determine whether an empirical indicator is a specific measure of status or may also be a measure of power.

Power as a structural capacity is independent of the intentions and expectations of actors. For example, take an exchange network in which three actors bargain for resources, $A1$—B—$A2$. Dashes signify potential exchange relations. Actor B can exchange with either $A1$ or $A2$ who cannot exchange with each other. We can create power in this network by restricting actors to a single exchange on any one round of bargaining. Actor B now has power because B has more potential partners than opportunities to exchange, while A actors must exchange with B to participate in the profits from exchange. Assuming that actors excluded from exchange receive no profit and that profit motivates actors, they will accept the best terms of exchange they can get. In order to avoid exclusion, actors in the A positions will bid against each other to offer terms more favorable to B. By accepting terms favorable to herself at the expense of another, B has used power. It is not necessary that B knows why the A actors are offering favorable exchange rates, nor even that B knows the exchange rate was favorable. The structure of the network confers power on B, and the favorable exchange rate is an indicator of that power. Actors' perceptions of power and related expectations may affect the relationship between status and power, but they do not affect power as a structural capacity.

Status, on the other hand, is conferred on a member by the group as a whole. Group members form expectations for each other's ability, and these expectations combine to produce a status hierarchy. The more consensus among group members concerning the ability assessments of group members, the greater the effect on the group's status hierarchy. In a situation where two group members disagree, no net effect on status rank should result. For example, both may have high expectations for their own ability and low expectations for each other's. In general, both high status and low status members reach a degree of agreement on each other's position in the group's status hierarchy. No such consensus is necessary for power.

THEORIES RELATING POWER TO STATUS

Relevant Exchange Theories

Emerson (1962) and Homans (1974) proposed different mechanisms relating power to status. Emerson's theory of power-dependence relations describes "status giving" as a balancing mechanism by which a low power actor can achieve a more equal footing with a high power actor. If a low power actor behaves in a deferential manner toward a high power actor, that actor is thought to become dependent on the low power actor for status. Consequently, the high power actor will be reluctant to jeopardize the relationship by continued overt power use. Thus power becomes balanced. For example, a boss's self-esteem may come to depend on the fawning behavior of subordinates. The boss will hesitate to discipline fawning employees fearing they might quit. This effectively neutralizes the boss's power. The "status giving" mechanism implies an asymmetry between high and low power actors in the way power produces status expectations. Only the low power actor infers a status differential. Emerson provides no mechanism by which high power actors might come to see themselves as high status or others as low status. For Homans (1974) also, power is more fundamental than status. Power that yields superior resources over time is inevitably represented by higher status. For example, a robber able to acquire money at gun point is not accorded high status because of his or her inability to use that power consistently. But a gangster able to demand "protection" money at will from merchants will eventually—that is, over time—be accorded higher status.

Relevant Expectation States Theories

In exchange theories, resources confer power that produces differences in rewards. *Reward expectations theory* (Berger et al. 1985) and Ridgeway's (1990) *resource expectations theory* relate reward or resource levels to expectations of ability or competence. In addition, high power and low power actors perform different roles. The *evaluation/expectation states theory* (Berger, Conner, and McKeown 1974) relates roles actors perform to expectations of ability. Expectation states theories link expectations of ability or competence to status rank. When combined, these theories relate power to status.

Power, Rewards, and Expectations

Berger, Fisek, Norman, and Wagner (1985) elaborated on ideas originated in Berger, Zelditch, Anderson, and Cohen (1972) concerning the formation of status expectations based on reward differences. Their theory assumes that actors share general beliefs about the correlation of individual and social attributes with reward levels in society at large. These are "taken for granted" beliefs that "everyone knows to be true." Termed "referential structures," these shared beliefs relate reward levels to categories of persons (such as race or gender), performance outcomes, or ability levels. If referential structures are activated, actors use them to form expectations for reward levels based on status characteristics, possessed relevant abilities, and performance outcomes in the immediate situation. This process may also work in reverse. For example in this society, people commonly believe that rewards are related to ability (the ability referential structure). Able people are supposed to be paid more than others. Thus the expectation may form that highly rewarded people are more able than others unless there is concrete evidence to the contrary.

For actors to relate reward levels to expectations for ability, referential structures must be activated in the local situation (Berger, Fisek, Norman, and Wagner 1985). Ridgeway and Berger (1986) argued that categorical referential structures are activated if a status characteristic is associated with the referential structure, and it discriminates between actors. Task situations usually engender concerns for reward equity. These concerns motivate actors to associate status characteristics with the ability referential structure. Therefore, in task situations where actors hold ability referential beliefs that discriminate between actors, and the situation does not explicitly prohibit such beliefs, Ridgeway and Berger assumed the ability referential structure is activated.

Once rewards are related to ability through the referential structure, actors ignorant of their own or others' ability may infer ability from knowledge of differences in rewards. This has received empirical support (Cook 1975; Harrod 1980; Bierhoff, Buck, and Klein 1986; Stewart and Moore 1992). This formulation relates exchange theories of power to status characteristics theory as follows: (a) Power differences allow actors to gain higher or lower rewards; (b) Actors form expectations for ability from different reward levels; and (c) Status rank is a function of expectations for ability.

Power, Resources, and Expectations

Ridgeway's (1990) theory also can be used to relate power to status. In exchange theory, power implies resources that actors may use to further group goals. These resources may be the source of power or may result from the accumulated rewards made possible by the use of power. Resources represent power in the sense of "power to" accomplish something (Wrong 1988) as opposed to "power over" another actor. Actors can use resources to influence events and further group goals. Resources are perceived as a kind of competence though independent of individual ability. For Ridgeway (1990, pp. 12–13), differences in resources convey the appearance of competence that is one component in the creation of status value:

> This appearance of competence should develop whether or not the resource rich individual is assumed to have "earned" or "deserved" his or her riches because of past valuable contributions or innate ability. Furthermore, this appearance of competence should develop whether or not the resource rich person ever actually uses his or her power in the situation, since it is the structural capacity to do so that gives the impression of competence.

Ridgeway's theory ties expectations of competence directly to resource levels. It is only necessary that differences in resource levels be known by group members for differences in competence assessments to appear. This relates exchange theories of power to status characteristics theory as follows: (a) Resources confer power. (b) Actors form expectations for competence from resources. (c) Status rank is a function of expectations for competence.

Power, Role Enactment, and Expectations

Another branch of expectation states theory offers a path by which power confers status. The evaluation/expectation states theory provides a mechanism by which a status hierarchy emerges in groups of status equals (Berger, Conner, and McKeown 1974; Fisek, Berger, and Norman 1991). This theory assumes that differences in performance expectations develop from differences in rates at which others accept performance outputs. (A performance output is a problem solving attempt by a group member.) An individual's role in a group is an external factor thought to contribute to the emergence of differences in group evaluations of individual performance attempts. Moore (1985) placed subjects in the role of a person rejecting the performance of a partner while accepting and staying with her or his own performance. In the second phase of the study subjects did reject the influence of the partner more often than did subjects who had not been placed in the role.

In many situations, a position of power affects the rate at which group members accept performance outputs. For example in an exchange situation, a high power actor with several alternatives for exchange seeks to reach agreement with a low power actor who has no alternative to exchange. The low power actor is more likely to accept an offer from the high power actor than the high power actor is to accept an offer from the low power actor. Accepting another's offer is a reward action that signifies a positive evaluation of the performance output. If this pattern of acceptance is repeated over time, it is comparable to the high power actor being placed in the role of a high status actor. This relates exchange theories of power to status characteristics theory as follows: (a) Acceptance rates for performance outputs emerge that are consistent with a position's level of power; (b) Actors form expectations for competence from acceptance rates; and (c) Status rank is a function of expectations for competence.

Theoretical Convergence: Power Confers Status

It is premature at this point to try to assess the relative strengths of the different theories discussed. Though these theories come from diverse perspectives, all reach the same conclusion. For Weber, as for Homans, power consistently applied over time confers status. Also, the theory relating power to status developed using expectation states theories reaches the same conclusion as do exchange theorists Homans (1974) and Emerson (1962). Emerson proposed that high power actors gain status when low power actors "give status" in an attempt to balance the situation. The expectation states formulations propose that expectations for actors' abilities form from knowledge of rewards, resources, and roles of actors in positions of power, and status rank is a direct function of these expectations of ability. Thus, as for Emerson, high power actors gain status. Analysis that begins with exchange theories converges on the same conclusion as analysis that begins with expectation states theories: a difference in power will produce a difference in status.

MEDIATORS BETWEEN POWER AND STATUS

The previous section showed how exchange, status value, and expectation states theories connect power to status. The question left to answer is, when do power differences not produce status differences? Both Ridgeway's (1984) work on dominance and Brehm's (1966) theory of *reactance* propose that actors may react negatively to overt power use. In addition, attributional biases may mediate in the process relating power to status.

Power Creates Resistance

Power, especially overt power use, can create resistance. It is possible that this resistance may prevent power from resulting in higher status for the power user in some situations. Emerson's (1962) balancing mechanisms reflect this idea (Samuel and Zelditch 1989). For Emerson, power use initiates a search by low power actors to find ways to counter it. For Brehm (1966; Brehm and Brehm 1981), low power actors see overt power use as an attempt to limit their freedom. In exchange terms, limiting freedom is equivalent to limiting the number of alternative courses of action open to the low power actor. Brehm proposed that attempts to limit freedom produce *reactance*. Reactance is an emotional reaction by low power actors that causes them to resist the influence attempts of high power actors. Wicklund and Brehm (1968) found that overt attempts to limit subjects' decision alternatives *negatively* influenced them compared to simple suggestions from their partners. That is, judgments tended to be in the direction opposite to that of the influence attempt. This result counters the preceding exchange and expectation states formulations. These predict that power use by high power actors will increase compliance by low power actors, even in situations where low power actors have nothing to lose by resistance.

Ridgeway's (1984) analysis of dominance attempts and status provides an explanation that reconciles these two apparently contradictory formulations. Ridgeway (1984) pointed out that not only is competence important to status expectations, but also *why* someone is perceived as competent. Expectations of competence produce increased influence only within the scope of expectation states theory. That is, group members must also believe that competence will be used to further group goals as opposed to purely individual goals. They are likely to consider dominance behavior self-oriented rather than group-oriented because of its competitive nature (Ridgeway 1978). "In most situations then, the competence indicated by dominance will probably be insufficient to outweigh the impression of self-interest" (Ridgeway 1984, p. 85). Dominance behavior may create the impression of competence but will not necessarily increase the influence of the actor attempting to dominate.

Empirical evidence is mixed. Wicklund and Brehm (1968) found that dominance attempts resulted in behavior opposed to the attempt. However, Driskell, Olmstead, and Salas (1993) conducted an experiment in which they crossed high and low states of task cues with high and low states of dominance behavior. Task cues resulted in increased influence but dominance behavior had no effect. The effect of dominance attempts on influence may depend on the power differences between actors. Power differences were not present in either study leaving the question open for investigation.

Dominance behavior is "power-oriented," based on threat of attack (Ridgeway 1984). It is plausible that actors interpret overt power use both as a dominance attempt and also as evidence of competence. It may be that a combining process occurs. The perception that the actor engaged in such an attempt is not group-oriented reduces influence that would have resulted from expectations of competence. This implies that power most effectively confers status when (a) overt power use is minimized and (b) group orientation is maximized.

Overt power users may still gain status from the exercise of their power but not directly from the group member exploited. Consider groups somewhat larger than a dyad. A powerful actor gains resources by dominating a few other group members. In a large group, members remain who may not be aware of this overt power use, or who may not interpret it as dominant behavior. These members may well see the powerful actor as group-motivated. For them, the resources acquired by the powerful actor will contribute to the powerful actor's increased influence. Overt power use results in an increase in status in this situation because the combined expectations of all group members produce status rank. In a large group, the positive expectations of many who only know of the resources powerful actors have may outweigh the negative expectations of the exploited few.

Knowledge of the Source of Power

Influence derived from expectations for competence may also depend on whether group members attribute competent performances to the actor or the situation. I have proposed that expectations of ability that actors form from different reward levels link status to power. But expectations for ability may depend not only on rewards but on inferences drawn from information concerning the source of rewards. For example, suppose you know that the bank balances of two millionaires are identical. You also know that one millionaire's wealth resulted from the exercise of options in the stock of the company she founded, and the other millionaire's wealth resulted from winning the lottery. Your expectations of ability for the company founder are likely to be higher than your expectations for the lottery winner. The empirical studies summarized here support the idea that the way actors process information affects the status derived from a power difference.

Four studies use behavioral measures of influence to test the effect of rewards on expectations of ability. In a study conducted by Joseph Berger and Hamit Fisek (reported in Cook 1970), experimental subjects had no information concerning relative ability levels except what they could infer from differential reward allocations assigned by lottery. They found some evidence that highly rewarded subjects developed greater expectations for their own ability than for their less well rewarded partners' ability in spite of being specifically informed of the randomness of rewards. However, an attempted replication by Webster in 1970–71 used the same explicit randomizing procedure and did not find a significant effect (Murray Webster, Jr., personal communication). Harrod (1980) conducted a similar experiment but did not tell subjects that differential rewards were randomly assigned. Instead, subjects believed that pay differences between themselves and their partners resulted from the standard practices of a research organization that had hired them. Highly rewarded subjects showed significantly higher expectations for their own ability than did less well rewarded subjects. Stewart and Moore (1992) replicated and elaborated on the Harrod study. Subjects randomly assigned to reward conditions were told their pay was determined by "information we have received from you." As in the Harrod study, highly rewarded subjects showed higher expectations for their own ability than did less well rewarded subjects.

Several studies that use self-report rather than behavioral indicators of status provide further evidence. Cook (1975) conducted a study of equity and expectations in which she manipulated inequity using differences in reward allocations assigned arbitrarily by "another student." In one of the conditions, differences in ability beliefs could only result from the inferences subjects drew from differences in reward allocation. Based solely on these differences, most subjects indicated on questionnaire items their belief that their own task ability was greater than their partners'. Whereas before the reward manipulation, all subjects reported believing that their own ability was equal to or less than that of their partners. Webster and Smith (1978) placed subjects in a condition that provided them with extremely low rewards compared to other subjects. Anecdotal evidence from post-experimental interviews suggests some subjects felt the size of their rewards might be due to their own lack of ability. Stolte (1983) manipulated structural power in an exchange situation that resulted in greater reward allocations for high power than for low power subjects. Subjects were unaware of the structure of the exchange situation that produced differences in power. High power subjects scored higher on post-test questionnaire measures of self-efficacy than did low power subjects. Cook, Hegtvedt, and Yamagishi (1988) manipulated structural power using a computer simulation to replace the subject's partner in exchange. On post-session questionnaire items, power advantaged subjects rated their own bargaining competence greater than their partners'. However, power disadvantaged subjects did not rate their own bargaining competence as lower than their partners'.

These studies taken together support the idea that information concerning the basis of differential reward allocations can decrease the status value of those rewards. Knowledge that rewards were randomly distributed seems to have reduced the effect of rewards. Stewart and Moore (1992) found a strong effect of reward difference on ability expectations while Berger and Fisek obtained only a small effect. The reason for this seems to be that in the Berger and Fisek study subjects possessed the knowledge that they were randomly assigned to high or low reward positions. While in the Stewart and Moore study, subjects were unaware of the real mechanism of assignment. Both Harrod's and Stewart and Moore's subjects had false information about the mechanism of assignment, either that subjects were assigned to pay conditions based on the "standard practices of the organization," or on "information we have received from you." Thus subjects may have inferred that their reward assignment was due to the researcher's knowledge of their abilities or status characteristics they possessed rather than from structural power in the situation. In fact in all three studies, subjects were randomly assigned to reward conditions. It was the structure of the situation, not subjects' abilities or status characteristics that determined subject pay. These studies support the idea that inferences made about the source of rewards can reduce the status conferred by power. However, an explanation of why this might be true requires some background in the attribution literature.

Attributional Biases and Expectation States

Attribution occurs when a person links an event to its possible causes. Heider (1944, 1958) analyzed how people explain the actions of others in everyday life, a process necessary to coordinating social interaction. He noted what may be the fundamental empirical generalization in the attribution literature. "Changes in the environment are almost always caused by acts of persons in combination with other factors. The tendency exists to ascribe the change entirely to persons" (Heider 1944, p. 361). Many attribution researchers have verified the existence of such a tendency.

Jones and Davis (1965) developed a theory of correspondent inferences:

> When the perceiver infers personal characteristics as a way of accounting for action, these personal characteristics may vary in the degree to which they correspond with the behavior they are intended to explain. Correspondence refers to the extent that the act and the underlying characteristic or attribute are similarly described by the inference (p. 223).

The tendency to attribute the cause of acts to the stable disposition of persons rather than to situational factors has come to be called *correspondence bias* (Jones 1986; Gilbert and Jones 1986).

In attribution research, ability is a stable disposition of an actor rather than a situational factor. It is at ability that correspondence bias connects with ideas of power and status. In expectation states theory, status rank is a direct function of expectations of ability. To the extent that actors make correspondent inferences of ability, expectations for ability result. To the extent that actors attribute the cause of performance to situational factors or luck, no expectations for ability result. Thus, correspondence bias may play a role in the production of expectations of ability and, therefore, status.

I found some support in the attribution literature. Jones and Goethals (1971) report evidence that actors tend to make attributions to ability early and persevere in them despite evidence to the contrary. Gilbert, Jones, and Pelham (1987) found that information about structural power can reduce correspondent inferences.

Self-serving attributional bias is also relevant to the relationship between power and status. People tend to attribute successful outcomes more to themselves—and to their stable dispositions such as ability—while attributing unsuccessful outcomes more to situational factors or luck (Pyszczynski and Greenberg 1987). Though some have argued against the existence of such an effect (Miller and Ross 1975), and one study finds the opposite of self-serving in a particular situation (Ross, Bierbrauer, and Polly 1974), Pyszczynski and Greenberg (1987) consider the finding well replicated. A search of the large literature on attribution turned up no studies on the interaction between self-serving and correspondence biases. However, it is possible to make some plausible assumptions about the nature of this interaction.

Support for attributional biases does not come from studies in which people choose exclusively *either* a dispositional *or* a situational attribution. Rather, the typical attribution study follows the experimental manipulation with a measure of whether the subject is *more or less* dispositional (e.g., Jones and Harris 1967; Quattrone 1982a). Other studies test for a *more or less* situational attribution (e.g., Quattrone 1982b).

In a prototypical attitude attribution experiment, Jones and Harris (1967) showed subjects an essay either praising or criticizing Fidel Castro then asked them to estimate the essay author's true attitude. Not surprisingly, subjects rated the authors of pro-Castro essays as more favorable to Castro than authors of anti-Castro essays. The evidence for dispositional bias comes from subjects' behavior in an experimental condition in which they were informed that author's were *assigned* to write either a pro-Castro or anti-Castro essay. This gave subjects specific evidence of a situational cause for the pro- or anti-Castro nature of the essay. Still, subjects who observed pro-Castro essays rated the author's true attitude to be more pro-Castro than did subjects who observed anti-Castro essays. This experiment produces no evidence that subjects rejected a situational cause in favor of a dispositional one. It does support the interpretation that a predisposition to dispositional attribution exists and, despite compelling evidence of a situational cause, dispositional attribution remains. In searching for evidence sufficient to satisfactorily explain an event, people may combine evidence for both dispositional and situational causes in making an attribution.

There is considerable support for the idea that people combine evidence of multiple causes of an event (Fleming and Darley 1989; Ginzel, Jones, and Swann 1987; Hamilton 1980; Johnson, Jemmott, and Pettigrew 1984; Jones and Goethals 1971; Jones, Riggs, and Quattrone 1979; Jones, Rock, Shaver, Goethals, and Ward 1968; Lord, Lepper, and Preston 1984; McArthur 1976; Ross, Lepper, and Hubbard 1975; Snyder and Jones 1974). They then may give greater weight to those that are dispositional and self-serving while discounting those that are situational or not self-serving in attempting to form some aggregate attribution of causation. Suppose the effects of self-serving and correspondence bias combine. What effect would this have on the relationship between power and status? Given the assumption that the two effects combine, I first develop a model using only correspondence bias then add the possible effects of self-serving bias to it.

Expectation states theories assume that to determine expectations of ability, group members combine all status information that has become relevant by the processes the theory describes. If power also produces expectations of ability, it seems reasonable that this assumption holds in this situation as well. Using the idea of correspondence bias, the above argument suggests that lacking complete information on all possible causes, participants will assume that a difference in reward outcomes is due to a personal disposition of the subject. That disposition is most likely to be ability. The ability referential structure has been activated because subjects are engaged in a task with their partners. In this society, rewards are strongly associated with ability. The "combining" assumption suggests that information on a situational cause will reduce the attribution to ability. But correspondence bias suggests that some attribution to ability will remain despite much evidence to the contrary. Even in the presence of knowledge that power results from a structural advantage unrelated to the ability of the actor, power conveys some status.

Figure 1 depicts the relative level of status expected to result from a specified level of power. Because of correspondence bias, actors assume that rewards *which in fact may be due to structural power* result from a disposition of the power holder such as ability or competence. Activation of the ability referential structure explains this link between performance and ability. Status rank is a function of expectations

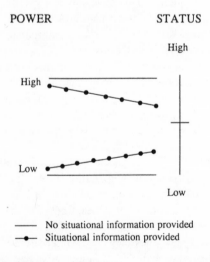

Figure 1. Effect of correspondence bias alone on status resulting from power

of ability or competence. Thus in the absence of information about the situational cause of power, solid lines run horizontally from high power to high status and from low power to low status. The dotted lines show the effect of providing information about situational cause. Correspondence bias continues to operate which means that high power still produces relatively high status. However, this effect is combined with the information about situational cause to reduce the status of the powerful. The dotted line from the low power actor is a mirror image of the one from the high power actor. Low power is still considered to result partially from lack of ability or competence on the part of the low power actor which results in relatively low status. However, information about situational cause raises the low power actor's status. Group members partially attribute the cause of the actor's low power to factors other than ability or competence. Providing information about the situational cause of power should reduce the effect of power on status.

Figure 2 depicts the expected effect of self-serving bias when combined with the effect of correspondence bias on the status derived from power. For both high power and low power actors, "self-serving" suggests that this bias will enhance the actor's self-expectations of ability, increasing her status. Thus, three of the lines in Figure 2 point to higher status positions than do comparable lines in Figure 1. However, for high power actors without situational information, the line is the same in both figures. High power actors without situational information assume their power results from their ability or competence. Self-serving bias does nothing to change this assessment.

POWER   STATUS

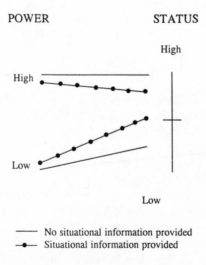

———— No situational information provided
—•— Situational information provided

Figure 2. Combined effect of correspondence *and*
self-serving biases on status

The effect of self-serving bias would be somewhat different on the aggregate ability expectations of a two-person group composed of a high power and low power actor. In accord with self-serving bias, a low power person will attribute her own failure and the success of a high power partner more to the situation. These situational attributions would diminish the low power actor's expectations of competence for her partner and increase them for herself. However, a high power person will make more dispositional attributions for her own success and her partner's failure. These dispositional attributions would increase a high power person's competence expectations for herself and decrease them for her partner. For the most part, the effects of self serving bias for high and low power members cancel each other leaving three of the lines in Figure 1 unchanged. However, the top line would slope down to the right. The status of a high power actor in the absence of information of situational causation would be reduced somewhat due to the situational attribution of the low power actor. In any case, the status rank of actors in the four conditions would remain the same after adding in possible effects of self-serving bias, but it does suggest that where power differences exist, consensus as to status rank will be harder to obtain.

The effect of correspondence bias remains. Actors will attribute power to the personal abilities of actors. However, actors given specific information that power results from social structural factors will make attributions that are less dispositional and more situational. Thus power produces differences in status, but those differences will be smaller when actors have specific information that the effects of power resulted from structural factors rather than the personal attributes of high power actors. Information that the effects of power resulted from structural (situational) factors counters but does not eliminate the correspondence bias that leads actors to attribute these effects to abilities (dispositions) of actors.

IMPLICATIONS

None of the theoretical formulations presented contradict the conclusion that power confers status. For some the relationship is all but inevitable. Both Weber and Homans suggested that in the long run, differences in power will produce differences in status. Ridgeway (1990) proposed that knowledge that actors possess different resource levels produces status differences regardless of how actors believe those resources were acquired.

Several formulations propose factors that may intervene to reduce status differences produced by power. Ridgeway's (1984) work on dominance and Brehm's theory of reactance suggest that actors and observers may react negatively to overt power use in ways that would reduce the influence of powerful actors. Research on attributions suggests that specific information about the structural source of power may reduce correspondence bias that leads actors to attribute powerful performances to the ability of actors rather than to situational factors. If power is linked to status through actors' expectations that rewards result from the dispositional abilities of actors, then specific information that rewards are due to structural power weakens this link and should reduce the power conferred by status.

Hypotheses

Hypotheses testable in a modest research program follow from my analysis. Design of these studies should be relatively straightforward, aided by the tradition of theory testing and standardized research settings available in both exchange and expectation states programs.

Weber and Homans proposed that power applied consistently over time confers status. Because influence is an indicator of status rank, this implies that the influence of a powerful actor over a weaker other will increase over time as power use continues.

Hypothesis 1. If an actor uses power repeatedly over an other, then the actor's influence over the other will increase.

Hypothesis 1, and all hypotheses that use influence as an indicator of status, should hold even when, after repeated power use, the situation changes so that the actor no longer has power over an other. An increase in status indicated by an increase in influence implies that actors' beliefs about the relative abilities of group members has changed. Once this change in belief occurs, it should remain though power differences no longer exist in the relationship. A typical status characteristics and expectation states experimental design could be used to test these hypotheses (e.g., Berger, Fisek, Norman, and Zelditch 1977).

Emerson proposes an asymmetry in the way that power confers status. Low power actors "give status" to high power actors to constrain their power use. Low power actors, by their behavior, suggest that high power actors have high ability relative to low power actors. Power does not directly produce increased assessments by a high power actor for her own ability. Low power actors should grant a high power other more influence than they would an equal power other. However, high power actors should not grant a low power other any less influence than they would an equal power other.

Hypothesis 2. If an actor is low in power, then a high power other will have more influence over the actor than would an equal power other; and, if an actor is high in power then a low power other will have no less influence than would an equal power other.

The analysis of attributional biases suggests that the amount of information available to actors about the structural sources of power has an effect on status differences that result. In the absence of information to the contrary, actors will attribute power differences to the abilities of actors. These ability expectations produce the status hierarchy. Specific information that power actually results from external factors, such as position in an exchange network awarded at random, should limit those ability expectations and reduce changes in status resulting from power differences.

Hypothesis 3.   Uninformed high power actors will have more influence over a low power other than will actors informed that power is due to the random assignment of actors to positions in an exchange network; and, uninformed low power actors will have less influence over a high power other than will informed actors.

Brehm's reactance theory proposes that power use creates a negative emotional reaction in low power actors that reduces or eliminates the influence of high power others. Hypotheses 4 and 5 follow.

Hypothesis 4.   A low power actor exchanging with a high power other will experience a more negative emotional reaction than will actors who exchange with an other equal in power.

Hypothesis 3 could be tested using a typical network exchange experimental design (e.g., Cook and Emerson 1978; Cook et al. 1983; Markovsky et al. 1988; Skvoretz and Willer 1991).

Hypothesis 5.   An actor will have less influence over an other who has experienced a negative emotional reaction toward the actor than would be the case in the absence of an emotional reaction.

Reward, resource and role expectations theories when linked with social exchange theories all propose that differences in power will produce differences in status rank. This implies that, *other things being equal*, differences in rewards, resources, and role behavior resulting from power differences will produce differences in influence. The phrase, "other things being equal," is important. It suggests that the effect should occur as long as emotional reactions remain constant for high and low power actors, whether emotional reactions are positive, negative, or neutral. That is, the effects of reward, resource, and role differences combine with the effect of emotional reactions to produce changes in the influence an other has on an actor.

Given a situation in which actors possesses either a high or low state of some status characteristic and also are either high or low in power, I make predictions about the relative influence of an actor based on the combination of status characteristic and power she or he possesses. For example, actors may be men or women if the status characteristic is gender, and these men and women may be either high or low in power relative to an other. Hypothesis 6 uses the combined effects of a status characteristic and power to predict the actor's influence. The effects of a status characteristic and power combine to order an actor's influence over an other as follows.

Hypothesis 6. A high power actor who also possesses the high state of the status characteristic relative to an other will have the most influence. A low power actor who also possesses the low state of the status characteristic relative to an other will have the least influence. Actors who have the high state of the status characteristic and low power relative to an other will have an intermediate level of influence as will actors who have the low state of the status characteristic and high power relative to an other.

For example in a typical network exchange situation, a low power actor's knowledge of an other's higher rewards would increase the other's influence. However, negative emotional reactions on the part of the low power actor would decrease the other's influence. If the effect of rewards and emotional reactions combine, then the influence of a high power other on a low power actor might remain the same as that of an other whose power equals the actor's.

CONCLUSION

The analysis of the relationship between power and status is complex. I found it necessary to cobble together theoretical concepts from several major areas of social psychology to describe that relationship. These included exchange theories of power, expectation states theories of status, Brehm's theory of reactance, and attribution research. Nonetheless, I will attempt to summarize the relationship between power and status in a sentence. High power actors gain status from their power to the extent that (a) power use does not arouse negative emotional reactions on the part of low power actors, and (b) information about the structural source of power is lacking, unclear or disregarded.

Specifying conditions under which power use produces more or less negative reactions is not only an empirical question but one that may require a theory from yet another major area of social psychology: emotions. This will add more complexity to the theoretical synthesis. It is a tribute to the maturity of social psychology as a discipline that theories from such diverse research areas can be fruitfully linked to approach a problem as fundamental as the relationship between power and status.

ACKNOWLEDGMENT

This work has benefited from financial support of National Science Foundation Grant #SES-9201124 and the advice of Morris Zelditch, Jr., Joseph Berger, and Cecilia Ridgeway.

NOTES

1. Several branches of the research program can be used to relate power to status including the *theory of reward expectations and status* (Berger et al. 1985) and the *status value theory* (Berger et al.

1972). These are covered in the section, Theories Relating Power to Status. Several works provide overviews of the program (Berger, Cohen, and Zelditch 1966, 1972; Berger et al. 1977; Webster and Driskell 1978; Berger et al. 1980; Berger et al. 1985; Webster and Foschi 1988) from which the ideas presented here have been culled.

2. This informal description is taken from Berger et al. (1980) and Berger, Fisek, and Norman (1989) where a formal description can also be found.

REFERENCES

Berger, J. 1958. "Relations Between Performance, Rewards, and Action-Opportunities in Small Groups." Unpublished Dissertation, Harvard University.

Berger, J., B. P. Cohen, and M. Zelditch, Jr. 1966. "Status Characteristics and Expectation States." Pp. 29–46 in *Sociological Theories in Progress*, Volume 1, edited by J. Berger, M. Zelditch, Jr., and Bo Anderson. Boston: Houghton-Mifflin.

———. 1972. "Status Characteristics and Social Interaction." *American Sociological Review* 37:241–255.

Berger, J. and T. L. Conner. 1974. "Performance Expectations and Behavior in Small Groups: A Revised Formulation." Pp. 85–109 in *Expectation States Theory: A Theoretical Research Program*, edited by J. Berger, T. L. Conner, and M. H. Fisek. Cambridge, MA: Winthrop.

Berger, J., T. L. Conner, and W. L. McKeown. 1974. "Evaluations and the Formation and Maintenance of Performance Expectations." Pp. 27–51 in *Expectation States Theory: A Theoretical Research Program*, edited by J. Berger, T. L. Conner, and M. H. Fisek. Cambridge, MA: Winthrop.

Berger, J. and M. H. Fisek. 1974. "A Generalization of the Theory of Status Characteristics and Expectation States." Pp. 163–205 in *Expectation States Theory: A Theoretical Research Program*, edited by J. Berger, T. L. Conner, and M. H. Fisek. Cambridge, MA: Winthrop.

Berger, J., M. H. Fisek, and R. Z. Norman. 1989. "The Evolution of Status Expectations: A Theoretical Extension." Pp. 100–130 in *Sociological Theories in Progress: New Formulations*, edited by J. Berger, M. Zelditch, Jr., and B. Anderson. Newbury Park, CA: Sage.

Berger, J., M. H. Fisek, R. Z. Norman, and D. G. Wagner. 1985. "Formation of Reward Expectations in Status Situations." Pp. 215–261 in *Status, Rewards, and Influence*, edited by J. Berger and M. Zelditch, Jr. San Francisco: Jossey-Bass.

Berger, J., M. H. Fisek, R. Z. Norman, and M. Zelditch, Jr. 1977. *Status Characteristics and Social Interaction: An Expectation States Approach*. New York: Elsevier.

Berger, J., S. J. Rosenholtz, and M. Zelditch, Jr. 1980. "Status Organizing Processes." *Annual Review of Sociology* 6:470–508.

Berger, J., D. G. Wagner, and M. Zelditch, Jr. 1985. "Introduction: Expectation States Theory: Review and Assessment." Pp. 1–72 in *Status, Rewards, and Influence*, edited by J. Berger and M. Zelditch, Jr. San Francisco: Jossey-Bass.

Berger, J. and M. Zelditch, Jr., eds. 1985. *Status, Rewards, and Influence*. San Francisco: Jossey-Bass.

Berger, J., M. Zelditch, Jr., B. Anderson, and B. P. Cohen. 1972. "Structural Aspects of Distributive Justice: A Status Value Formulation." Pp. 119–146 in *Sociological Theories in Progress*, Volume 2, edited by J. Berger, M. Zelditch, Jr., and B. Anderson. Boston: Houghton-Mifflin.

Bierhoff, H. W., E. Buck, and R. Klein. 1986. "Social Context and Perceived Justice." Pp. 165–185 in *Justice in Social Relations*, edited by H. W. Bierhoff, R. L. Cohen, and J. Greenberg. New York: Plenum.

Blau, P. M. 1964. *Exchange and Power in Social Life*. New York: John Wiley and Sons.

Brehm, J. W. 1966. *A Theory of Psychological Reactance*. New York: Academic Press.

Brehm, S. S. and J. W. Brehm. 1981. *Psychological Reactance: A Theory of Freedom and Control*. New York: Academic Press.

Cohen, E. G., R. Lotan, and L. Catanzarite. 1988. "Can Expectations for Competence be Altered in the Classroom?" Pp. 27–54 in *Status Generalization: New Theory and Research*, edited by M. Webster, Jr. and M. Foschi. Stanford, CA: Stanford University.

Cohen, E. G. and S. Roper. 1972. "Modification of Interracial Interaction Disability: An Application of Status Characteristics Theory." *American Sociological Review* 37:643–657.

———. 1985. "Modification of Interracial Interaction Disability." Pp. 350–378 in *Status Rewards and Influence*, edited by J. Berger and M. Zelditch, Jr. San Francisco: Jossey-Bass.

Cook, K. S. 1970. "Analysis of a Distributive Justice Experiment: Goal Objects and Task Performance Expectations." Unpublished, Stanford University.

———. 1975. "Expectations, Evaluations and Equity." *American Sociological Review* 40:372–388.

Cook, K. S. and R. M. Emerson. 1978. "Power, Equity and Commitment in Exchange Networks." *American Sociological Review* 43:721–739.

Cook, K. S., R. Emerson, M. R. Gillmore, and T. Yamagishi. 1983. "The Distribution of Power in Exchange Networks: Theory and Experimental Results." *American Journal of Sociology* 89:275–305.

Cook, K. S., K. A. Hegtvedt, and T. Yamagishi. 1988. "Structural Inequality, Legitimation, and Reactions to Inequity in Exchange Networks." Pp. 291–308 in *Status Generalization: New Theory and Research*, edited by M. Webster, Jr. and M. Foschi. Stanford, CA: Stanford University.

Driskell, J. E., B. Olmstead, and E. Salas. 1993. "Task Cues, Dominance Cues, and Influence in Task Groups." *Journal of Applied Psychology* 78:51–60.

Emerson, R. M. 1962. "Power-Dependence Relations." *American Sociological Review* 27:31–40.

———. 1964. "Power-Dependence Relations: Two Experiments." *Sociometry* 27:282–298.

———. 1969. "Operant Psychology and Exchange Theory." Pp. 379–405 in *Behavioral Sociology*, edited by R. L. Burgess and D. Bushnell, Jr. New York: Columbia University.

———. 1972. "Exchange Theory Part II: Exchange Relations and Network Structures." Pp. 58–87 in *Sociological Theories in Progress*, Volume 2, edited by J. Berger, M. Zelditch, Jr., and B. Anderson. Boston: Houghton-Mifflin.

Festinger, L. 1953. "An Analysis of Compliant Behavior." Pp. 232–256 in *Group Relations at the Crossroads*, edited by M. Sherif and M. O. Wilson. New York: Harper Brothers.

Fisek, M. H., J. Berger, and R. Z. Norman. 1991. "Participation in Heterogeneous and Homogeneous Groups: A Theoretical Integration." *American Journal of Sociology* 97:114–142.

Fleming, J. H. and J. M. Darley. 1989. "Perceiving Choice and Constraint: The Effects of Contextual and Behavioral Cues on Attitude Attribution." *Journal of Personality and Social Psychology* 56:27–40.

Gerth, H. H. and C. W. Mills. 1958. *From Max Weber: Essays in Sociology*. New York: Galaxy.

Gilbert, D. E. and E. E. Jones. 1986. "Perceiver-Induced Constraint: Interpretations of Self-Generated Reality." *Journal of Personality and Social Psychology* 50:269–280.

Gilbert, D. T., E. E. Jones, and B. W. Pelham. 1987. "Influence and Inference: What the Active Perceiver Overlooks." *Journal of Personality and Social Psychology* 52:861–870.

Ginzel, L. E., E. E. Jones, and W. B. Swann. 1987. "How 'Naive' is the Naive Attributor?: Discounting and Augmenting in Attitude Attribution." *Social Cognition* 5:108–130.

Hamilton, V. L. 1980. "Intuitive Psychologist or Intuitive Lawyer? Alternative Models of the Attribution Process." *Journal of Personality and Social Psychology* 39:767–772.

Harrod, W. J. 1980. "Expectations from Unequal Rewards." *Social Psychology Quarterly* 43:126–130.

Heider, F. 1944. "Social Perception and Phenomenal Causality." *Psychological Review* 51:358–374.

———. 1958. *The Psychology of Interpersonal Relations*. New York: Wiley.

Homans, G. C. 1974. *Social Behavior: Its Elementary Forms*. New York: Harcourt Brace Jovanovich.

Johnson, J. T., J. B. Jemmott III, and T. F. Pettigrew. 1984. "Causal Attribution and Dispositional Inference: Evidence of Inconsistent Judgments." *Journal of Experimental Social Psychology* 20:567–585.

Jones, E. E. 1986. "Interpreting Interpersonal Behavior: The Effects of Expectancies." *Science* 234:41–46.

Jones, E. E. and K. E. Davis. 1965. "From Acts to Dispositions: The Attribution Process in Person Perception." *Advances in Experimental Social Psychology* 2:219–266.

Jones, E. E. and G. R. Goethals. 1971. "Order Effects in Impression Formation: Attribution Context and the Nature of the Entity." Pp. 27–46 in *Attribution: Perceiving the Causes of Behavior*, edited by E. E. Jones, D. E. Kanouse, H. H. Kelly, R. E. Nisbett, S. Valins, and B. Weiner. Morristown, NJ: General Learning Press.

Jones, E. E. and V. A. Harris. 1967. "The Attribution of Attitudes." *Journal of Experimental Social Psychology* 3:1–24.

Jones, E. E., J. M. Riggs, and G. Quattrone. 1979. "Observer Bias in the Attitude Attribution Paradigm: Effect of Time and Information Order." *Journal of Personality and Social Psychology* 37:1230–1238.

Jones, E. E., L. Rock, K. G. Shaver, G. R. Goethals, and L. M. Ward. 1968. "Pattern of Performance and Ability Attribution: An Unexpected Primacy Effect." *Journal of Personality and Social Psychology* 10:317–340.

Kemper, T. D. and R. Collins. 1990. "Dimensions of Microinteraction." *American Journal of Sociology* 96:32–68.

Lenski, G. 1966. *Power and Privilege: A Theory of Social Stratification*. New York: McGraw–Hill.

Lord, C. G., M. R. Lepper, and E. Preston. 1984. "Considering the Opposite: A Corrective Strategy for Social Judgment." *Journal of Personality and Social Psychology* 47:1231–1243.

Markovsky, B., D. Willer, and T. Patton. 1988. "Power Relations in Exchange Networks." *American Sociological Review* 53:220–236.

McArthur, L. Z. 1976. "The Lesser Influence of Consensus than Distinctiveness Information on Causal Attributions: A Test of the Person-Thing Hypothesis." *Journal of Personality and Social Psychology* 33:733–742.

Miller, D. T. and M. Ross. 1975. "Self-Serving Biases in the Attribution of Causality: Fact or Fiction?" *Psychological Bulletin* 82:213–225.

Molm, L. D. 1981. "The Conversion of Power Imbalance to Power Use." *Social Psychology Quarterly* 44:151–163.

———. 1985. "Gender and Power Use: An Experimental Analysis of Behavior and Perceptions." *Social Psychology Quarterly* 48:285–300.

———. 1990. "The Dynamics of Power in Social Exchange." *American Sociological Review* 55:427–447.

Moore, J. C., Jr. 1985. "Role Enactment and Self Identity." Pp. 262–315 in *Status, Rewards and Influence: How Expectations Organize Behavior*, edited by J. Berger, D. G. Wagner, and M. Zelditch, Jr. San Francisco: Jossey-Bass.

Pugh, M. D. and R. Wahrman. 1983. "Neutralizing Sexism in Mixed-Sex Groups: Do Women have to be Better than Men?" *American Journal of Sociology* 88:746–762.

Pyszczynski, T. and J. Greenberg. 1987. "Toward an Integration of Cognitive and Motivational Perspectives on Social Inference: A Biased Hypothesis-Testing Model." *Advances in Experimental Social Psychology* 20:297–340.

Quattrone, G. A. 1982a. "Behavioral Consequences of Attributional Bias." *Social Cognition* 1:358–378.

———. 1982b. "Overattribution and Unit Formation: When Behavior Engulfs the Person." *Journal of Personality and Social Psychology* 42:593–607.

Ridgeway, C. 1978. "Conformity, Group-Oriented Motivation, and Status Attainment in Small Groups." *Social Psychology Quarterly* 41:175–188.

———. 1981. "Nonconformity, Competence, and Influence in Groups: A Test of Two Theories." *American Sociological Review* 46:333–347.

———. 1984. "Dominance, Performance, and Status in Groups: A Theoretical Analysis." Pp. 59–93 in *Advances in Group Processes*, Volume 1, edited by E. J. Lawler. Greenwich, CT: JAI Press.

———. 1990. *"The Social Construction of Status Value: Gender and Other Nominal Characteristics."* Presented to the 85th Annual Meeting of the American Sociological Association. Washington, DC, August 11–15.

Ridgeway, C. and J. Berger. 1986. "Expectations, Legitimation, and Dominance Behavior in Task Groups." *American Sociological Review* 51:603–617.

Ross, L. D., G. Bierbrauer, and S. Polly. 1974. "Attribution of Educational Outcomes by Professional and Nonprofessional Instructors." *Journal of Personality and Social Psychology Bulletin* 29:609–618.

Ross, L. D., M. R. Lepper, and M. Hubbard. 1975. "Perseverance in Self-Perception and Social Perception: Biased Attributional Processes in the Debriefing Paradigm." *Journal of Personality and Social Psychology* 32:880–892.

Samuel, Y. and M. Zelditch, Jr. 1989. "Expectations, Shared Awareness, and Power." Pp. 288–312 in *Sociological Theories in Progress: New Formulations*, edited by J. Berger, M. Zelditch, Jr., and B. Anderson. Newbury Park, CA: Sage.

Skvoretz, J. and D. Willer. 1991. "Power in Exchange Networks: Setting and Structural Variations." *Social Psychology Quarterly* 54:224–238.

Snyder, M. L. and E. E. Jones. 1974. "Attitude Attribution when Behavior is Constrained." *Journal of Experimental Social Psychology* 10:585–600.

Stewart, P. A. and J. C. Moore, Jr. 1992. "Wage Disparities and Performance Expectations." *Social Psychology Quarterly* 55:78–85.

Stolte, J. F. 1983. "The Legitimation of Structural Equality: Reformulation and Test of the Self-Evaluation Argument." *American Sociological Review* 48:331–342.

Thibaut, J. W. and H. H. Kelly. 1959. *The Social Psychology of Groups.* New York: J. Wiley and Sons.

Webster, M., Jr. and J. E. Driskell, Jr. 1978. "Status Generalization: A Review and Some New Data." *American Sociological Review* 43:220–236.

Webster M., Jr. and M. Foschi, eds. 1988. *Status Generalization.* Stanford, CA: Stanford University.

Webster, M., Jr. and L. F. Smith. 1978. "Justice and Revolutionary Coalitions: A Test of Two Theories." *American Journal of Sociology* 84:267–292.

Wicklund, R. A. and J. W. Brehm. 1968. "Attitude Change as a Function of Felt Competence and Threat to Attitudinal Freedom." *Journal of Experimental Social Psychology* 4:64–75.

Wrong, D. R. 1988. *Power: Its Forms, Bases, and Uses.* Chicago: University of Chicago.

A NEW THEORY OF
GROUP SOLIDARITY

Barry Markovsky and Edward J. Lawler

ABSTRACT

This paper examines previous conceptualizations of group solidarity and related concepts in the sociological and social psychological literatures. After identifying ambiguities in previous usages, we define solidarity in terms of relational patterns among actors. Specifically, solidarity is defined in terms of two network properties: the relative directness of ties among actors, and the homogeneity of those ties. In other words, solidarity exists for a given set of actors to the degree that they are directly connected to each other and there is an absence of subgroups or cliques. Although a variety of relational bases are conceivable, we illustrate our new conceptualization in a theory of emotion-based group solidarity. We further develop our formulation to account for the emergence of emotional bonds to the group *as a group* and the impact of vicarious experiences on emotional processes.

Advances in Group Processes, Volume 11, pages 113–137.
ISBN: 1-55938-857-9

Our goal in this chapter is to analyze the idea of "group solidarity" and provide an alternative conceptualization that sidesteps problems in earlier usages. Two general literatures provide the background for our analysis: group solidarity and group cohesion. The solidarity literature consists primarily of theoretical analyses by sociologists (e.g., Durkheim 1956; Gamson, Fireman, and Rytina 1982; Hechter 1987; Coleman 1990), and the cohesion literature consists primarily of empirical analyses by psychologists (e.g., Festinger, Schachter, and Back 1950; Cartwright and Zander 1968; Forsyth 1990). These distinct literatures contain some common themes that are essential to our own effort, but both lack the conceptual rigor necessary for the development of cumulative theory. In fact, a critical problem that pervades both of these literatures is the lack of any clear and consistent distinction between the nature of group solidarity (or cohesion) per se and its ostensible determinants and consequences. We begin with an overview of each body of literature, highlighting ideas that will be useful in the subsequent presentation of our reconceptualization.

BACKGROUND

The Solidarity Literature

The study of solidarity has a long history in sociology, beginning with Emile Durkheim's (1956 [1893]) distinction between mechanical and organic forms. Mechanical solidarity occurs in groups that contain people who are similar in background, values, and beliefs. There is an emotion-based sense of community in such groups, and the norms that are part of the community constitute a strong force constraining individuals. In other words, there is a strong and specific "collective conscience" that enhances uniformity of behavior across individuals. Organic solidarity, in contrast, develops out of differences rooted in divisions of labor. Durkheim indicated that as the division of labor and concomitant specialization grows, the interdependence of society's parts (e.g., individuals and the positions they occupy) becomes the primary foundation for social solidarity. Interdependence, however, produces a weaker collective conscience than mechanical solidarity and, relatedly, less individual uniformity and compliance.

In characterizing the types of groups in which mechanical or organic solidarity occurs, Durkheim (1956) sought necessary conditions for the development of each form of solidarity, for example, "similar members" is a necessary condition for mechanical solidarity, "specialization" is necessary for organic solidarity. These distinctions have proven useful by identifying two independent and complementary dimensions underlying group solidarity. However, Durkheim did not provide an unambiguous definition of solidarity, distinct from the determinants he discussed. We will return to this problem shortly.

Durkheim's distinction between organic and mechanical solidarity corresponds generally with two alternative approaches to social order found in contemporary sociological literature: utilitarian and emotional. The former begins with the assumption that social order is created and maintained because (and only if) interdependence makes cooperation a valued commodity. People join and remain in groups because groups provide rewards that are not available elsewhere. Groups then must develop systems of monitoring and sanctioning to maintain sufficient compliance to group norms and prevent members from "free-riding." Thus, the utilitarian approach views groups both as sources of reward and as agents of behavioral constraint.

The utilitarian approach to solidarity is illustrated in the recent work of Hechter 1984, 1987) and Coleman (1990). Hechter cites (1) the extensiveness of corporate obligations, and (2) the probability that members fully comply with these obligations as "defining properties" (1987, p. 18) of solidarity, but also indicates that solidarity is a "joint product" of these factors (1987, p. 52). The extensiveness of corporate obligations is determined by the dependence of members on the group, and the degree of compliance with those obligations is given by the degree to which the group monitors and sanctions the behavior of members. Lacking monitoring and sanctioning, solidarity would be broken by the actions of "free-riders." Thus, dependence on the group and monitoring are the primary determinants of solidarity. Hechter does not clearly define solidarity independent of these determinants.[1] He does assert, however, that "solidarity varies with the proportion of members' private resources that are contributed to collective ends" (1987, p. 168), suggesting at least one operational definition that is distinct from dependence and monitoring.

Coleman's (1990) theory is similar to Hechter's in its general approach and in certain of its details. (See Turner 1992 for analyses and an integration of the two theories.) Arguments in his theory are much more formalized than Hechter's, however, articulating as it does with an earlier mathematized theory (e.g., Coleman 1986). Briefly, that theory bases various social phenomena on patterns of actors' interests in goods over which others have control, and in those actors' levels of control over goods in which others have interests. The system coheres to the degree that these interests result in exchange relations and, further, to the degree that actors act upon their interest in controlling free-riding. A "second-order free rider problem" must then be solved: it is in the rational actor's interest to both promote norms and sanctioning systems against free-riding, but also to avoid contributing to it. If all did so, then the normative and sanctioning systems would collapse. An interesting solution that Coleman proposes is an escalation into zealotry, whereby actors develop prescriptive norms and informal systems of mutual rewards for desired behaviors. The result of this cycle of positive action and reward is that group members become zealots in pursuit of a common goal, and the need for second-order monitoring and sanctioning is eliminated. A caveat is that such zealotry is only possible in densely connected networks.

Hechter's theory provides further specificity to Durkheim's organic (or utilitar ian) dimension. His approach recognizes that the ties which bind members to group may be rooted in their dependence on the group for valued resources, and tha monitoring and sanctioning systems contribute to the maintenance of group viabil ity and the fulfillment of obligations. Coleman's (1990) theory does not focus o group solidarity per se (and thus sheds no new light on the definitional issue), bu its problem focus is nevertheless virtually identical to Hechter's. Exchange rela tions are maintained through time with the aid of behavioral sanctions and interes induced dependence. Unlike Hechter, however, under certain conditions Coleman theory allows for the possibility of a reward-based, as opposed to a strictl cost-based, sanctioning system.

The determinants specified by Hechter and Coleman seem quite reasonable t us, at least as *sufficient conditions*, even if we are still working with only an implici notion of group solidarity. However, perhaps due to a bit of zealotry of their owr both theorists seem to implicitly treat their determinants as the exclusive routes t solidarity, that is, as sufficient *and necessary* conditions. In contrast, we assert tha a clearer definition—one that more clearly demarcates determinants from definin properties—will facilitate the opening of theoretical routes to group solidarity othe than those based on material exchange. For example, solidarity may be charac terized by relations of dependence, compliance, or threat of negative sanction, o by bonds enacted in the spirit of voluntarism (as distinct from zealotry), *or*, as w shall see, by other means.

In contrast to the utilitarian approaches, the "emotional" approach to solidarit owes more debt to Durkheim's notion of mechanical solidarity. It begins with th assumption that social order is created and maintained by the affective ties o individuals to groups. People join and stay in groups because of their emotiona attachments to groups and their members. From this general standpoint, grou activities foster positive sentiments, enjoyable relations, and a sense of belongin that becomes objectified in collective objects, symbols, rituals and the like (Durk heim 1956; Berger and Luckmann 1966; Collins 1981; Lawler 1992).

Although they incorporate some utilitarian notions, Gamson and his associate exemplify the emotional approach by focusing on the more socio-emotional sid of group memberships. Specifically, Fireman and Gamson (1979, pp. 21–22 identified five factors that constitute the basis for a "person's solidarity with group": the member (1) has friends and relatives that are in the group, (2) act collectively with other members, (3) has his "design for living" shared and sup ported by other members, (4) shares with other members the same set of subordinat and superordinate relations with outsiders, and (5) believes that exit from the grou would be difficult. These group ties generate a sense of common identity, share fate, and general commitment to defend the group. Solidarity is essentially a produc of a collective identity and the strength of interpersonal ties (Gamson, Fireman, an Rytina 1982, p. 22).

A problem with this approach is the relative lack of both rigorous definitions of concepts—especially solidarity—and explicit statements of assumptions. It is not clear which, if any, of the bases of solidarity are primary, and which are assumed to work through those primary determinants. Finally, it is unclear whether Gamson and his associates consider solidarity to be an emergent property of groups, or whether it consists of the distribution of individual sentiments. Using the phrase "a person's solidarity with the group" (Fireman and Gamson 1979, p. 21) may suggest the latter, distinguishing this approach from others which view solidarity as an emergent, group level property. Overall, the work of Gamson and associates, though insightful, reveals conceptual difficulties similar to those found in the utilitarian approach.

Parsons' (1951) analysis of instrumental and expressive action suggests a useful distinction for understanding both the solidarity and cohesion literature. Specifically, he contrasted the relationship of members to each other in a group (e.g., interpersonal ties) and the relation of members to the collectivity as an object of perception (e.g., person-collectivity ties). He portrayed solidarity as institutionalized loyalty involving these two types of relations. First, the cathectic-expressive integration of ego with alter, where alter serves as the source of an organized system of gratifications, and the relation is embedded in a system of established role relations; second, the ego-collectivity relation which he describes as follows:

> By extension of this conception of expressive loyalty between individual actors we derive the further important concept of the loyalty of the individual actor to a collectivity of which he is a member. The collectivity may be treated as an object of attachment . . . it is clearly the collectivity, not its members as individuals, which is the significant object (1951, pp. 77–78).

Thus, the collectivity is itself an object in Parsons' formulation.

Parsons' (1951) distinction between the collectivity and other group members as objects of attachment seems crucial for understanding solidarity. The attachment to the collectivity as an object, distinct from its particular members, may distinguish groups with high solidarity from those with low solidarity, although such person-to-group ties may ring hollow without some degree of interaction among members of the collectivity. The institutionalized role structure makes the collectivity salient to actors (i.e., objectifies the collectivity) and facilitates interpersonal ties during role enactments. These micro ties typically develop within multiple, small subsets of group members.

More recently, Scheff (1990, p. 201) has defined solidarity in terms of social "attunement." In his words, attunement is "mutual understanding; joint attention to thoughts, feelings, intentions, and motives between individuals but also between groups" (p. 199) and "long-range considerations involving intention and character" (p. 201). By definition, a social bond exists between people to the degree that there is mental and emotional attunement between them, and "the same kind of attunement between groups is referred to as solidarity" (p. 201). Scheff's original thesis mainly focuses on an emotion-based process assumed to create and maintain bonds. This is tantamount to an explanation of social *cohesion*, as defined here, whereas solidarity appears to be defined as a simple aggregation of bonded or attuned actors. Scheff's greatest potential contribution is in his conceptualization of the social bond-formation process. His conceptualization of solidarity per se does not add significantly to prior work, nor was it so intended. However, to the degree that one wishes to pursue the study of emotion-based paths to group solidarity, Scheff's micro-interaction theory may be well worth considering.

To summarize our review of group solidarity literature, we find that despite the evident interest and obvious centrality of the solidarity concept to sociologists, the amount of research is vanishingly small. One looks in vain for a cumulative, ever-improving body of theoretical and empirical knowledge about group solidarity. We believe this failure is attributable to a lack of emphasis on developing precise, testable theories. Though general ideas abound, researchers have had few explicit propositions to test and no systematic framework for connecting the many ad hoc hypotheses derivable from the rather loose theorizing about solidarity. The cohesion literature, though more empirically oriented, reveals somewhat similar problems.

The Cohesion Literature

Since the early work of Festinger and associates (Festinger et al. 1950), cohesion has been a central focus of research in the group-dynamics tradition of psychology (e.g., Gross and Martin 1952; Eisman 1957; Lott and Lott 1965; Shaw 1981; Kellerman 1981; Ridgeway 1983; Drescher, Burlingame, and Fuhriman 1985). Nearly all of the work in this tradition starts from the Festinger, Schachter, and Back (1950) field-theoretical definition of cohesion as *the sum total of forces which act on members to remain in the group*. Others have offered more specific definitions that stress particular "forces," for example, the attraction of group members to each other, the attraction of members to the group (i.e., to its activities, goals, or leaders), the motivation of members to remain in the group, and the degree to which the group mediates valued outcomes (Gross and Martin 1952; Eisman 1957; Piper, Marrache, Lacroix, Richardson, and Jones 1983; Evans and Jarvis 1980; Drescher et al. 1985). In some instances, authors have intended these more specific forces to be operational indicators of the larger concept and, in others, they have proposed substituting the more specific concept or variable for the original one (Gross and Martin 1952). In any case, the most frequent conceptual or operational indicator of cohesion has been the level of interpersonal attraction, often measured by sociometric choices (Lott and Lott 1965; Gross and Martin 1952; Evans and Jarvis 1980; Shaw 1981).

Psychologists have conducted much more research on cohesion than sociologists have on solidarity, and so there is an extensive empirical literature on the consequences of cohesion (see Drescher et al.'s 1985 review). Varied definitions and operationalizations prevent firm conclusions, but some implications may be drawn (e.g., Shaw 1981; Ridgeway 1983; and Drescher et al. 1985). For example, there is evidence that groups with higher cohesion produce greater interaction among members, more compliance to group norms, higher satisfaction with the group, and lower levels of absenteeism (Festinger et al. 1950; Keller 1983). The link between cohesion and productivity is more complex, but groups with higher cohesion are generally more effective at achieving whatever goals the group adopts (Goodacre 1951; Norris and Niebuhr 1980), whether this entails higher or lower levels of productivity in a work setting (Seashore 1954). Extreme levels of cohesion, however, may impair decision processes by producing "groupthink" (Janis 1972) or "deindividuation" (Zimbardo 1969).

Determinants of cohesion have been shown to include attitude similarity (Newcomb 1961), intergroup tension (Sherif and Sherif 1953; Tajfel and Turner 1979), self disclosure by members (Stokes, Feuhrer, and Childs 1983), and the degree to which the group meets its members' needs (Shaw 1981; Ridgeway 1983; Nixon 1979; Palazzolo 1981). Despite the accumulation of empirical findings, however, cohesion research has shown minimal theoretical development over time. It is more the case that lists of factors that have operated as determinants and consequences in various empirical settings continue to grow, while cohesion theories fail to benefit in any systematic fashion.

Several conceptual and measurement problems have plagued this literature, many traceable to the classic work of Festinger, Schachter, and Back (1950). From the original formulation, one could argue either that cohesion is a multidimensional phenomenon capturing the field of forces acting on individuals, or that cohesion is the resulting impact of these forces on group members (see especially Gross and Martin 1952). Early evidence indicated that various standard measures of cohesion are not highly correlated (Eisman 1957), but later work has not been able to resolve the multidimensionality problem, theoretically or empirically (see Feldman 1968, 1973; Piper, Marrache, Lacroix, Richardson, and Jones 1983; Drescher et al. 1985). Efforts to develop a unidimensional concept or unitary measure of cohesion as a result of the force field have tended to settle on the attraction of members to the group or some variant of this (Eisman 1957; Piper et al. 1983).

Recent arguments suggest a distinction between cohesion (as bonds tying individuals together) and attraction to the group (Evans and Jarvis 1980). For instance, these dimensions were identified in a factor-analytic study by Hagstrom and Selvin (1965). It is noteworthy that the basic distinction between the interpersonal and person-group dimensions of solidarity have a parallel in the cohesion literature. Overall, however, it is hardly an understatement to say that within this literature there is little agreement or uniformity in any respect. The issues debated in the early 1950s are of the most basic sort, and it is arguable whether any of them have been resolved.

The two literatures most relevant to this paper—solidarity and cohesion—reveal complementary problems. The solidarity literature contains broad conceptualizations and discursive analyses that have not led to testable theories (Hechter 1987 and Coleman 1990 are exceptions). The cohesion literature consists of a set of disparate empirical findings guided by a loose conceptualization of the phenomenon and ad hoc hypotheses. The fundamental need of both literatures is a theory or set of theories that foster cumulative theoretical and empirical work.

Without claiming to solve all the problems of these literatures, we first present an abstract and general definition for the concept of group solidarity, exploring some of its properties and implications. We then provide a theory of group solidarity that explicates the role of emotions in the formation, maintenance, and dissolution of solidary groups. Rather than casting a loose, broad net designed to capture all previous work on solidarity and cohesion, our purpose is to devise the smallest possible set of concepts and propositions that will enable the most general possible explanation of solidarity. The definition of solidarity is designed to be useful for a variety of theoretical bases, for example, emotional, rational, structural, and so forth. Our theory will look in but one of these possible directions.

THE CONCEPTS OF COHESION AND SOLIDARITY

One of our major points of contention regarding previous work on solidarity and cohesion is the lack of a clear definition for these concepts. In our own struggle with this problem, we have discovered what we believe to be a reason for this: both concepts carry with them numerous intuitively appealing connotations, most of which we readily associate with cohesion, solidarity, or both. This is probably because these connotations are factors that we readily conceive of as causes or consequences. This is the same trap that most of the cited research and theories fell into. Thus, we began by working on definitions that would sidestep the distracting connotations of previous conceptions and theories, highlighting not what leads to or follows from solidarity, but rather the distinctive properties by which it can be identified.

In his contribution to Kellerman's (1981) volume on group cohesion, Sudarshan wrote that:

> . . . cohesion is the causal framework for structure and form. Structure implies parts which go together to make a whole and a search for the bonding agency leads us to study cohesion. By its very nature, then, cohesion implies constraint to freedom, but rather than destroy freedom altogether it tames the tendency to chaos and channels it into harmony (1981, p. 124).

Although its bearing on our interests is direct and obvious, it so happens that Sudarshan is a physicist and he was writing about the cohesiveness of physical matter. Also writing in Kellerman's collection, another physicist suggested that the cohesion of a set of objects is increased by factors that reduce the chaos of relations among members (Silber 1981). He noted that the states of matter (liquid, solid, gas) are properties of groups of atoms that are predictable from the properties of the individual atoms and their relations, but not exhibited by the individual atoms. The emergent properties of a substance in a particular state are determined by the relative degree of order (or absence of chaos) in interatomic and/or intermolecular relations. That is to say, macro properties are determined by the organization or structuring of components.

"Structure" implies a degree of stability in the spatial relationships of components. As noted in any basic physics text (e.g., Feynman, Leighton, and Sands 1963), the state of matter occupied by a substance depends in part on the balance of attractive and repulsive forces between particles. At very small distances a strong repulsive force prevents atoms and molecules from collapsing in on themselves. At relatively larger distances, particles attract one another. The actual distances involved depend on the nature of the substance, the degree to which its atoms or molecules are "excited" (i.e., its temperature), and other factors.

It is the relative stability of the inter-particle distances that determine the state of matter—the degree of structure—exhibited by the set of particles. Gases, characterized by extreme chaos, possess a near total lack of structure. Liquids are more structured in that temporary bonds among subsets of particles form and, as a consequence, the collection of particles coheres to some degree. Solids are even more structured in that atoms or molecules remain in fixed relationships to one another, their motions restricted to "vibrating" in place. All of a substance's characteristic emergent properties—the temperature at which it changes states, its density, its brittleness if a solid, its viscosity if a liquid, its conductivity, and so on—depend on properties of the constituent particles and their relations to one another.

These physical insights regarding the relation of parts to the cohesive whole suggest a general theoretical strategy: to understand solidarity and cohesion, we should consider (1) a small set of relevant properties of individuals comprising the group, (2) the nature of the relations between group members, and (3) properties of the structure of relations characterizing the group. Within these guidelines, we propose abstract structural definitions for sets of human actors that are analogous to the states of matter. As a scope condition, we will initially consider sets of uniform and simple actors, each of whom may engage in relations with one or more other actors in the set.[2]

We begin with a definition of cohesion. Our definition relies on the concept of "reachability" from social network analysis. Reachability refers to the strength and directness of relations among members of a set of actors. It is maximized when every actor has a strong, direct relation with every other actor in the set. Reachability is reduced when some actors are only indirectly related (e.g., A and C in the $A—B—C$ network), and/or when some relations are weaker. We may thus define cohesion as follows:

Definition 1. *A set of actors is cohesive to the degree that it has high reachability.*

Note that this definition does not specify any particular type of relation. Simply, a relation is any sort of attractive bond between actors, and thus a relation from x to y ("$x{\to}y$") indicates only and precisely that *something* attracts x to y. For instance, x may be attracted to y because y is physically appealing, y is attracted to x, y is a friend of a friend, or because x sees a mutual relationship with y as potentially remunerative, fulfilling, comforting, useful, and so forth. The specific factors that make relations operate as "bonding agents" are probably countless. One may choose to pursue whatever line of theorizing one chooses, and we would expect that different relational bases will require different theoretical tacks. These may relate cohesion to cognitive, behavioral or other properties of individual actors, to properties of certain combinations of different types of actors, to certain structural configurations, external forces, and so on. Our emotion-based theory, presented here, explores one such theoretical direction.

We next define three distinct pure types of "actor sets," analogous to the three states of matter. We call these aggregates, assemblages, and solidary groups. As with physical matter, natural sets of actors that constitute pure instances of any given type will be exceedingly rare, perhaps even nonexistent. However, the pure forms serve as excellent points of reference, and are thus worth defining explicitly.

Aggregates, in the first definition, are analogous to gases. Relations in a pure aggregate are weak and chaotic if they exist at all, and exhibit no discernable structure. In short:

Definition 2.   *An aggregate is a set of actors with low cohesion.*

The concept of aggregate serves as a baseline to which more organized sets of actors may be compared. The paradigm case is a set of strangers waiting at a bus stop. They are contained by the vessel of a common purpose—to board the same bus at the same time—but lack any further social structuring with respect to one another.

Let us next make explicit the intuition that the durability of a relationship is given by its strength. A set of actors whose members have generally weak relations may then form substructures whose shapes change over time, or which stand in relatively "fluid" relations to one another. A pure assemblage is thus a perfectly "liquid" group whose integrity is maintained not via a fixed structure, but by the "stickiness" that results from properties of the distribution of relatively *temporary* relations among its members. Thus, relative to the low cohesion of an aggregate:

Definition 3.   *An assemblage is a set of actors with moderate cohesion.*

Given that cohesion is defined in terms of reachability, we may then say that every pair of actors in an assemblage is directly or indirectly related, but not necessarily at all times. The spectators at a football team's home games is such a grouping. Across a season, "membership" in the assemblage is fluid. At gatherings, relationships among members are often locally strong (people come with friends) but always globally weak. Despite the low average strength of relationships, however, it is sufficient for various forms of collective action, for example, making a "wave." In fact, the wave illustrates very well the low but non-zero reachability among actors in the assemblage: spectators in *Section A* can "reach" those in *Section M* via a chain of influence stretching from *A* to *B* to *C* and so on.

Moderate cohesion at the level of the assemblage can also result from fluctuations in the strengths of a fixed set of relations. Thus, at a given time, flows of influence, resources, communication, sentiment, and so forth, may be relatively concentrated in some relations, then in other relations at other times. An example may be found in shifting patterns of coalition building and busting among members of a political body, responding to series of issues to be decided collectively.

It is pleasantly fortuitous that the word "solid" denotes not only the third state of matter, but also appears in the label for the third type of set, the solidary group:

Definition 4. *A solidary group is a set of actors with high cohesion and unity of structure.*

High cohesion again implies that members are in relatively strong, enduring and direct relations with one another. "Unity of structure" is the absence of substructures, that is, subsets of actors with higher reachability among one another than with those in other subsets. It may be that the paradigm case of a cohesive group is one having all actors strongly related to all others, thus producing a structure that also has perfect unity. In general, however, two groups may be similar in terms of reachability, but one configuration may be more unitary than the other. For example, both sets of actors in Figure 1 have six members, six strong relations, and two weak relations. The configuration of relations in 1a, however, clearly shows two strong substructures linked only by weaker relations. In contrast, the location of the weak relations in 1b does not result in the formation of distinct substructures. Therefore, 1b is a more solidary group. Another way of viewing the distinction between solidarity and cohesion is that the former imposes certain higher-level structural conditions, such as structural homogeneity.

The question now is not whether we have captured some "essential" property of solidarity, some essential distinction between it and cohesion, or what some particular set of theorists think of when they think about solidary groups. Rather, we are interested in whether the definitions and distinctions that we have drawn will aid in the development of a parsimonious and fruitful theory. Toward this consideration, we next examine some of the implications and properties associated with our definitions.

Properties of Pure States

If the different states of cohesion—aggregate, assemblage, solidary group—are to be useful concepts in a theory, they should at the very least suggest a unique set of properties for each state. Table 1 illustrates such properties. We will examine

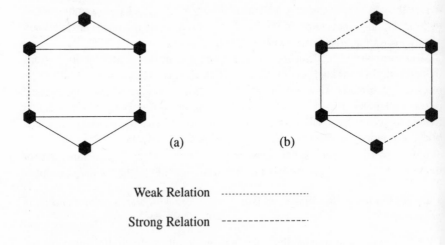

(a)   (b)

Weak Relation   ----------------------

Strong Relation -----------------

Figure 1. Equally cohesive groups that vary in solidarity

each row of the table in sequence. Note that we consider these notions provisional and somewhat speculative, but interesting nonetheless.

1. *Strength.* All else being equal, a higher degree of cohesion implies greater structural strength. That is, to the extent that a given type of relation is strong enough to hold a pair of actors together within some larger context or environment, then a larger set of actors held together by the same type of bonds within the same

Table 1. Properties of Aggregates, Assemblages and Solidary Groups

Property	Form		
	Aggregate	Assemblage	Solidary Group
Strength	low	moderate	high
Stability	low	moderate	high
Largest Discernable Unit	actor	cluster	group
Largest Enduring Unit	actor	actor	group
Locus of Chaotic Activity	inter-actor	inter-actor, inter-cluster	none
Boundary Definition	arbitrary	variable	fixed
Penetrability	high	moderate	low

type of environment will also resist cleavage. By extension, then, solidary groups are stronger (in this sense) than assemblages which, in turn, are stronger than aggregates.

2. *Stability.* The increase in stability as we move from aggregate to assemblage to solidary group is entailed by the way we characterized reachability. Reachability implies strength, strength implies endurance, and the greater the endurance of pair-wise relations, the greater the endurance of the larger configurations they comprise.

3. *Largest Discernable Unit.* If we take a "time-slice" of the structure of relations among a set of actors, we would find disjoint actors in the aggregate, disjoint clusters of actors in the assemblage, and no disjointedness in the solidary group.

4. *Largest Enduring Unit.* If instead of a time-slice we examine persistence over time, only in the solidary group would any enduring non-actor unit be discernable: the group as a whole.

5. *Locus of Chaotic Activity.* In aggregates, actors behave chaotically with respect to one another. In assemblages, actors may form temporary clusters, but the clusters behave chaotically with respect to one another. Theoretically, a solidary group is devoid of such chaotic activity at any level, although it may behave chaotically with respect to other groups, clusters, or actors.

6. *Boundary Definition.* Any boundary describing an aggregate is arbitrary or artificial, that is, not inherent in the aggregate itself. In contrast, assemblages have sufficient cohesion that the boundary discriminating the assemblage from its environment should be discernable—analogous to the effects of surface tension of liquids. The changing relations in assemblages, however, result in different clusters and different actors forming the boundary at various times. The unitary structure of the pure solidary group, in contrast, produces a stable boundary.

7. *Penetrability.* It is conceivable that under some conditions, the relative penetrability of a set of actors is related to its level of internal cohesion. The changeable relations in assemblages (or absence of relations in aggregates) could more readily permit the intrusion of new actors, in contrast to the case of the relatively strong and stable relations of the more solidary group. An exception may occur in a solidary group with formal positions. If an actor leaves a position, such groups may have a structural imperative to fill their gaps. A nonmember actor with the appropriate characteristics should be attracted to the gap by the same forces that attracted the former occupant. Even still, the imposition of selection criteria in effect inhibits the flow of actors into such a group.

Degrees of Solidarity

To solve the problem of defining solidarity, we had to back-track and define a more fundamental property, cohesion, to which solidarity is related but not identical. If solidarity requires a strong, unitary relational structure, are there degrees of solidarity? Here again the physical analogy provides some insight. A 12-inch cube of iron is relatively solid with respect to most earthly environments. However, the identical material in many such environments will be much more brittle when its gross (as opposed to fine) structure is altered in certain ways, for example, when it is formed into a five-foot square sheet only one-half inch thick. In some respects, such a change can make the object even more solid, for example, with respect to a blow to its half-inch side. In most respects, however, the solidity is weakened and the object is much more vulnerable to breakage. The analogy to social structural change would be a shift from a situation in which most people have strong ties to most others, to one in which strong ties are arranged relatively linearly. The reduction in solidarity is then attributable to the decreased average reachability between members.

Another way that the solidity of physical objects can be altered is through the introduction of "impurities." This is not always the case, as iron is strengthened into steel by "doping" it with carbon. Often, however, impurities alter gross properties of the substance into which they are introduced. In its pure form, for example, water does not gradually harden as it solidifies, nor gradually soften as it melts. It makes no sense to talk of water ice that is less than solid. In many impure substances and more complex compounds, however, state changes are not so discrete. Butter softens before it melts. Glass has certain liquid properties even in its "solid" form. Changes in temperature can transform rubber compounds through liquid, semi-solid, and solid states with no definite points of transition.

We may conceive of group structures in an analogous way. For example, a group could have an open, lattice-like structure making it "solid" but "brittle," or members may possess a capacity to form new relations when old ones are broken due to external pressures, imparting a degree of "resilience" to the structure. In groups with different types of relational ties—"impure" groups, so to speak—we would expect that factors that serve to weaken ties of a certain type do not necessarily weaken ties of other types. Such selective tie-weakening may shift a set of actors from a state of solidarity to something stronger than an assemblage—from solid to "slushy" rather than liquid. In general, it makes sense to presume that in natural groups with complex, nonuniform actors, solidarity is usefully conceptualized as a continuum. The problem now is to demonstrate the theoretical utility of such a conceptualization.

EMOTIONS AND SOLIDARITY

For purposes of sociological analysis, emotions are individual phenomena with the potential for affecting group-level social change. One of the ways this can happen is through the weakening or strengthening of bonds between group members in response to changes in the emotional states of those members. Emotions may then be construed as one type of "glue" that can bind people to one another and to their groups, thereby maintaining the coherence and integrity of supra-individual social units. Emotions are not the only basis for such bonds, but they clearly are a significant one. If we can show how the emotions of individuals can create, modify, and destroy social structures, or how the response of individuals to group processes are conditioned or mediated by emotional processes, we will have made a significant advance toward understanding how emotions and emotional processes relate to group solidarity. The specific purpose of this section is to theoretically examine the role of emotion in the creation and maintenance of group solidarity. We assert that under certain conditions emotions delimit, if not determine, emergent group structures.

Working now at the higher end of the cohesion spectrum, we will consider the role of emotions in the production of group solidarity. Our definition for solidary groups requires unity of structure and high reachability. Reachability, in turn, is based upon the strength and directness of relations. Thus, anything that can be assumed to act on assemblages so as to increase the strength of relations, or the directness of relations, and/or the unity of the structure, without reducing one of the other factors, is therefore sufficient to increase group solidarity.

Our approach to solidarity thus directs us toward, among other things, relational bases that bond the members of sets of actors. Although numerous "bonding agents" may be identified for complex human actors, here we limit our attention to emotions. Thus we now expand the scope of our theory to include sets of actors, the members of which are capable of anticipating and experiencing a desirable positive emotion such as excitement, joy, wonder, love, or contentment, or an undesirable negative emotion such as fear, anger, despair, hatred, or frustration. The actual emotional experience of an actor x stems from one or more behavioral encounters with another actor, y.

Emotion/Solidarity Propositions

Our theory picks up from the point at which an actor, x, anticipates experiencing a positive emotion through an encounter with a chosen member, s_i, of a set of actors, S. The theory makes no assumptions about the source of x's expectation, nor the reasons that encounters may engender a particular emotional response. In addition to these scope conditions we reiterate that (for now) the hypothetical actors to which this theory refers are essentially uniform and very simple.[3]

Our first proposition links x's expectations with subsequent action:

Proposition 1. If x expects to experience positive emotions in an encounter with s_i, then x will encounter s_i.

Note also that for this and the other propositions, negative emotions entail opposite consequences, for example, in this case if x anticipates negative emotions, then x will not encounter s_i.

The second proposition establishes the connection between emotions and the formation of relations:

Proposition 2. If x encounters s_i and x experiences positive emotions, then $x \rightarrow s_i$ strengthens.

With reference to the physical metaphor, positive emotions enhance the attractive forces between members, negative emotions diminish them. In combination, Propositions 1 and 2 suggest that groups that induce expectations of positive emotional encounters will produce high rates of interaction if those expectations are generally confirmed.

The third proposition posits a tendency for reciprocation to occur in pair-wise relations:

Proposition 3. If $x \rightarrow s_i$ strengthens, then $s_i \rightarrow x$ strengthens.

In natural groups, it has been found that reciprocity of attractive relations is commonplace (Berscheid 1985) although the actual strengths of the mutual relations are generally not perfectly matched (Wellman 1988). Our theory would predict such differentials in strengths, given the next proposition.

The fourth proposition captures the idea that all else being equal, if an actor has positive experiences with one member of a set of actors, then he or she will anticipate similar experiences with other members. This provides the mechanism for a process of emotional generalization:

Proposition 4. If $x \rightarrow s_i$ strengthens, then x anticipates experiencing positive emotions in an encounter with an s_j.

When "fed back" into the other propositions, it becomes clear that if x's expectation is verified, the $x \leftrightarrow s_i$ relation will form and strengthen with further positive encounters. Thus, assuming positive encounters, the strengthening of relations among some members will strengthen relations among others.

The foregoing propositions provide an elementary emotion-based explanation for the emergence of a structurally defined group solidarity. If all of the interpersonal encounters in S result in positive emotional responses, perfect solidarity will emerge in S *whether or not each member interacts with every other member*. If some encounters produce mixed emotions and weaker relations, considerable solidarity may still emerge, depending on the structural location of those weaker relations. In fact, all three structural requisites for solidarity—strength, directness, and unity—are achievable via the processes described by the propositions. Propositions 2 and 3 involve strength-enhancement, and Propositions 3 and 4 together establish processes that increase the directness of relations and unity of structures.

These are basic ideas indicating how a unitary structure with high reachability might develop among a set of actors. The potential utility of our framework is revealed by extending these propositions to other phenomena associated with solidarity. For example, both solidarity and cohesion literatures suggest in a loose way that the group or collectivity somehow becomes an object in itself for members (Parsons 1951; Hagstrom and Selvin 1965; Piper et al. 1983), and an extension of these propositions offers a specification of this process. In addition, we can indicate how something like a "resistance to identity loss" might create repulsive forces that prevent groups from reaching extreme levels of solidarity, and also how vicarious experiences in groups strengthen solidarity further. The following pages briefly analyze these three processes in order to illustrate the potential value of the elementary propositions above.

Emotional Bonding to Groups

Let us define a concept g_x representing total strength of all $x \leftrightarrow s_i$, that is, of x's direct, mutual relations. Consider what can happen if x is allowed to develop a somewhat different kind of emotional relation.

Proposition 5.   The greater g_x, the stronger the positive emotions x associates with S.

The greater the frequency and intensity of relations between x and members of the group, the stronger the positive emotional bonds to the group as an object. This specifies a process by which interpersonal relations generate emotional bonds that connect members to the group.

Interestingly, emotional bonding to the group can vary independently from group solidarity since it is based only on direct ties among actors. It then becomes possible to conceive of groups whose members all feel strong emotional ties to the group, while the group has relatively low solidarity; all actors are overgeneralizing their emotional responses and actually reside in disparate emotional subgroups. There is research that shows that people do tend to make inferences about groups on the basis of insufficient samples (e.g., Quattrone and Jones 1982; Tversky and Kahneman 1971). In such cases, these structures should be far more vulnerable to disintegrative forces than those having higher structural unity and reachability.

On the other hand, it may be reasonable to propose that emotional attachment to the group may facilitate structural solidarity by fostering positive encounters with previously unrelated others. Perhaps once a threshold of g_x is exceeded, x develops such a strong bond with the group that he or she becomes unconditionally attracted to other members whom he or she encounters.

Resistance to Identity Loss

Thus far we have focused on attractive forces that maintain relationships. Earlier, however, it was stated that particles comprising physical matter are also governed by a repulsive force at short distances that prevents matter from collapsing in upon itself. An analog for our set of emotion-experiencing actors could be a negative emotional reaction to the loss of individual identity. That is, as an actor becomes increasingly enmeshed in a network via strong attractive relations, he or she may suffer a loss of individuality. The actor becomes more a structural element than an actor, losing individual integrity and certain freedoms (Zimbardo 1969). Resistance to such losses may be conceptualized as an increase in x's repulsive force toward others, limiting the strength and/or number of relations in which x may engage. Let t_x represent x's threshold in S, above which relationships become identity-robbing. The effect of resistance to identity loss is given as:

Proposition 6.   The greater $g_x - t_x$, the greater the degree to which x experiences negative emotions in one or more $x \rightarrow s_i$.

Using earlier propositions we can derive that:

Derivation 1.   *The greater $g_x - t_x$, the greater the degree to which one or more* $x \rightarrow s_i$ weakens.

When t_x is exceeded in S but x still wishes to establish new relations, he or she may seek them with members of other sets of actors. In this manner multiple group memberships may be established. Regardless of x's relations with other sets of actors, however, coupling Proposition 6 with those given previously creates a countervailing force against solidarity. The result may be shifting clusters of related actors in S—the analog to a "liquid group." Whether S is an aggregate, assembly or solidary group will depend on the actual levels of t_x for members of S. In general, then, resistance to identity loss is one process that can limit potentially deleterious consequences of extreme solidarity such as deindividuation and groupthink.

Vicarious Responses

What would happen if actors' emotional responses were affected not only by their direct encounters, but also by the encounters of those to whom they are related? One simple and well-grounded way to introduce this possibility is to allow our actors to (1) make inferences about other relations, (2) observe whether those inferences are confirmed or disconfirmed, and (3) respond emotionally to this observation.

We first propose that actors will tend to infer consistency among relations. In particular:

Proposition 7.   If $x \leftrightarrow s_i$ and $x \leftrightarrow s_j$ are strong, then x infers that $s_i \leftrightarrow s_j$ is strong.

Then,

Proposition 8. If x infers that $s_i \leftrightarrow s_j$ is strong and observes that s_i experiences positive emotions in an encounter with s_j, then x experiences positive emotions.

In short, x is assumed to be capable of vicarious emotional responses to social inconsistencies. Cooper and Fazio (1984), Croyle and Cooper (1983), Elkin and Leippe (1986), and Fazio and Cooper (1983) all provide evidence for "dissonance arousal," a physiological response to cognitive inconsistencies. Batson and Coke (1981), Berger (1962), Bandura and Rosenthal (1966), and Krebs (1975) provide evidence for empathetically induced physiological/emotional experiences. The propositions suggest that x takes an interest in the goings-on of those to whom he or she is strongly attached. If we allow that x's observing an s_i—s_j encounter itself qualifies as an encounter for x, then x's emotional response is predicted by Proposition 2. Combined with subsequent propositions, this will lead x to anticipate positive encounters with other members of S and possibly establish new relations. Thus, a relation in which x is not a member may affect x's emotional state and, as a result, affect the number and strength of x's relations. In this manner, members may come to perceive a commonality and mutuality of experience in the group.

The true nature of the s_i—s_j relation would only be a good predictor of x's vicarious response if x perceived the encounter accurately, or if s_i and s_j accurately project their emotional responses to the encounter. It is possible for x to misperceive the encounter as positive, to anticipate positive encounters with other members of S, to then engage in encounters with others, form new relations, and so forth. In this way a single misperceived relation can have an impact far beyond and opposite that warranted by its true nature.

IMPLICATIONS AND CONCLUSIONS

This chapter is part of a larger theoretical effort to develop a structural theory of group solidarity (see Markovsky and Chaffee 1995). It focuses on sets of actors whose members have the potential to form and maintain relations that may vary in their strength and directness. Cohesion is thereby conceptualized in structural terms and, in conjunction with structural considerations, viewed as a partial determinant of group solidarity.

The explanatory power of a theory depends as much on the comprehensibility of its terms as on the assertions it makes. With this in mind, one purpose of this chapter has been to develop clearer concepts of cohesion and solidarity, thereby laying the foundation for examining the effects of emotional responses (and other factors) on group solidarity. The theory focuses on the processes whereby positive or negative emotional responses in dyadic encounters affect the strength of relationships, and how larger patterns of relationships determine the degree of group cohesion or solidarity. We have made no formal claims about the causes of emotions nor the consequences of the solidarity that may follow.

Given the conceptual problems and related issues in the cohesion and solidarity literatures, it is important to understand how our approach bears on these literatures. The classic definition of cohesion is open-ended, that is, a field of forces that motivate individuals to remain in the group. The definition fails to distinguish the causes of cohesion (i.e., the forces) from cohesion itself (i.e., remaining in the group), and this has generated substantial confusion and debate over the years (Gross and Martin 1952; Piper et al. 1983). In these terms, our formulation treats cohesion as an *effect* of the "field of forces" and defines the nature of that effect as the development of relations among a set of actors. The dimensions that distinguish groups with lower vs. higher levels of cohesion are the strength and endurance of the relations, and the degree of reachability among actors. The key question, then, is how do strong, enduring, and reachable relations develop among a set of actors?

We have shown that one pathway to strength and stability is through positive emotions that develop from encounters between actors. Our theory indicates that such positive emotions undergird attraction between parties to encounters, that under certain conditions such emotions spread to other relationships, and that they may even create emotional bonds to the group as an object unto itself. In this sense, our theory systematically disentangles and interrelates the fundamental phenomena encompassed by Parsons' (1951) distinction between alter and the group as an object, Hagstrom and Selvin's (1965) distinction between "social satisfaction" and "sociometric cohesion," and Evans and Jarvis' (1980) distinction between "cohesion" and "attachment-to-group." Our approach would take issue with the need to develop an elaborate multidimensional conceptualization of cohesion or solidarity (see the recent effort of Drescher et al. 1985) but would certainly indicate a multidimensional conception of the causes of solidarity. Emotion is but one of many possible determinants.

It may prove instructive to consider how emotions could mediate some of the more commonly cited determinants of cohesion or solidarity such as the dependence of the actors on the group (e.g., Hechter 1987; Shaw 1981), intergroup conflict (Tajfel and Turner 1979), and sanctioning systems (Hechter 1987).

The dependence of members on the group motivates encounters and exchanges with other group members (Emerson 1972). If these encounters are rewarding, we would expect them to produce positive emotional responses. Thus, even groups appearing to rest on utilitarian foundations could generate emotional relations among members (Tallman, Gray, and Leik 1991). Utilitarian approaches to solidarity then may overlook the true causes of group solidarity, such as the bonds between group members. It is possible that the only condition under which dependence or other utilitarian factors could produce and maintain high solidarity without emotional relations is when actors have no alternative groups and no prospect of developing alternative sources of gratification, that is, when members' dependence on the group is absolute. Even then, however, there is no assurance of interpersonal bonding. The solidarity may be minimal.

The effect of intergroup conflict resembles that of dependence. The task of dealing with hostile others fosters intragroup encounters. The successful coordination of efforts, and especially the successful resolution of the conflict, would strengthen the positive emotions associated with intragroup relations and bridge subgroups that may previously have been only loosely related. The initial level of cohesion or solidarity would then be enhanced by the intergroup hostility.

Another utilitarian tenet is that people remain in groups that provide rewards, and/or levy costs for leaving. Groups typically create sanctioning and selective incentive mechanisms to promote member compliance to formal and informal regulations and the accomplishment of group tasks. Here again, the emotional attachment of members should play a crucial mediating role. A social control agency may use either positive or negative sanctions (rewards or punishments) to obtain particular types of compliant behavior. The emotional implications of these two types of sanctions are obviously very different. Groups that attempt to control members through the administration of positive rather than negative sanctions are more likely to produce positive emotional responses and thereby foster attractive bonds among members, and between members and the group. Thus, positive sanctions promote cohesion and solidarity. Groups that punish members for noncompliance may succeed in controlling aspects of members' behaviors, but the negative emotional responses associated with punishment should weaken member-group relations and possibly even inter-member relations. Exceptions to the latter may occur if members are capable of providing one another with emotional support and/or rewards that countervail the punishments meted by social control agents.

Our theory helps to provide more rigorous derivations of some of the consequences of solidarity suggested in related literatures. For example, the close relations that are required for high reachability should facilitate communication and mutual influence among group members. Also, the heightened emotional attachment of members to the group should both increase voluntary compliance to group norms and decrease free-riding (see Lawler 1992). Examples of possible group level effects are the development of formal institutions that monitor the behavior of individuals, that systematize the application of positive and negative sanctions, and that regulate access to and exit from the group. Thus, human groups can develop the capacity to purposefully manipulate and modify their own emergent properties, unlike the states of matter.

A social network imagery has informed our theorizing in many ways. Several concepts, such as relation, reachability, structural completion, boundary, and structural unity, reflect this imagery. In addition, a network metaphor has inspired us to think in multilevel terms, and informed our effort to connect individuals' emotions to the development of a set of stable relations (i.e., a social structure). Our passage from physical metaphors to an actual social theory was facilitated, if not explicitly guided, by general notions about social networks.

One of our next tasks is to formalize our theory using some of the network tools that have become available in recent years (e.g., Burt 1982, 1987; Burt and Minor

1983; Wellman and Berkowitz 1988). Its scope may then be broadened to include multiple types of relations among actors and multiple subgroups, permitting a broader range of derivations regarding macrolevel phenomena. Already our propositions hint of mathematical expressions that may be extrapolated, tested, and made even more precise. Additional structural concepts could also be incorporated and their implications drawn out. We have scarcely begun to consider, for example, how structural positions and variations in relational strength may interact to affect solidarity, the possible effect of simultaneous memberships in multiple networks, and the role of weak ties between subgroups within the group (Granovetter 1973, 1982). While there is much territory between emotions and group structure that remains to be explored, penetrating deeply into this territory is unlikely to lead anywhere without a conceptual and theoretical foundation such as that proposed in this paper.

ACKNOWLEDGMENTS

We are grateful for the thoughtful comments on earlier drafts of this paper that were provided by Randall Collins, Anne Eisenberg, Karen Heimer, Jodi O'Brien, Guy L. Siebold, Lynn Smith-Lovin, Jonathan Turner, participants of the Iowa Workshop on Theoretical Analysis, and graduate students in the first author's 1993 Seminar on Group Solidarity.

NOTES

1. By explicitly defining solidarity in the same terms that he asserts to be its determinants, Hechter presents an unusually bald tautology. The problem this creates is critical: solidarity cannot exist unless its "determinants" are present *by definition*. This makes the asserted causal relationships logically untestable. Obviously, Hechter did intend dependence and monitoring to be determinants and not defining properties. Failing to provide those defining properties, however, leaves those who may wish to test it with no guidelines for developing empirical indicators for "group solidarity." Hechter then loses any control over how other researchers operationalize solidarity for purposes of testing *his* theory.

2. See Lawler, Ridgeway, and Markovsky (1993) for a related general strategy for building theories in "structural social psychology."

3. Some readers of earlier drafts of this chapter have noted a correspondence between the assumptions to follow and Homans' (1974) "basic propositions." They are worth paraphrasing here: (1) the frequency of an act is determined by the frequency of reward; (2) an act is more likely to the degree that an associated stimulus is similar to one associated with the act and a prior reward; (3) the likelihood of an act is determined by the value of its result; (4) each additional reward of a given type has diminishing value; and (5) approving or aggressive behavior follows from, respectively, receiving or not receiving an expected reward. These are fairly standard behavioristic assumptions and as such, capable of serving as interpretations of virtually *any* behavior. Some of the propositions that we assert could be interpretable through, or reducible to, Homans'

basic propositions. However, they are not logically derivable from Homans' propositions because they make assertions with respect to entities and phenomena that Homans' did not address, that is, expectations, relations, encounters, relational strength, bonding to groups, and so on.

REFERENCES

Bandura, A. and T. L. Rosenthal. 1966. "Vicarious Classical Conditioning as a Function of Arousal Level." *Journal of Personality and Social Psychology* 3:54–62.

Batson, C. D., and J. S. Coke. 1981. "Empathy: A Source of Altruistic Motivation for Helping." In *Altruism and Helping Behavior*, edited by J. P. Rushton and R. M. Sorrentino. Hillsdale, NJ: Earlbaum.

Berger, P. L. and T. Luckmann. 1966. *The Social Construction of Reality*. New York: Doubleday.

Berger, S. M. 1962. "Conditioning Through Vicarious Instigation." *Psychological Review* 69:450–466.

Berscheid, E. 1985. "Interpersonal Attraction." In *The Handbook of Social Psychology*, Volume II, 3rd ed., edited by G. Lindzey and E. Aronson. New York: Random House.

Burt, R. S. 1982. *Toward a Structural Theory of Action*. New York: Academic Press.

———. 1987. "Social Contagion and Innovation: Cohesion Versus Structural Equivalence." *American Journal of Sociology* 92:1287–1335.

Burt, R. S. and M. J. Minor, eds. 1983. *Applied Network Analysis: A Methodological Introduction*. Beverly Hills, CA: Sage.

Cartwright, D. and A. Zander. 1968. *Group Dynamics: Research and Theory*, 3rd ed. New York: Harper and Row.

Coleman, J. S. 1986. *Individual Interests and Collective Action*. New York: Free Press.

———. 1990. *Foundations of Social Theory*. Cambridge, MA: Harvard University Press.

Collins, R. 1981. "On the Microfoundations of Macrosociology." *American Journal of Sociology* 86:984–1014.

Cooper, J. and R. H. Fazio. 1984. "A New Look at Dissonance Theory." Pp. 229–266 in *Advances in Experimental Social Psychology*, Volume 17, edited by L. Berkowitz. New York: Academic Press.

Croyle, R. T. and J. Cooper. 1983. "Dissonance Arousal: Physiological Evidence." *Journal of Personality and Social Psychology* 45:782–791.

Drescher, S., G. Burlingame, and A. Fuhriman. 1985. "Cohesion: An Odyssey in Empirical Understanding." *Small Group Behavior* 16:3–30.

Durkheim, E. 1956. *The Division of Labor in Society*. New York: Free Press.

Eisman, B. 1957. "Some Operational Measures of Cohesiveness and their Interrelations." *Human Relations* 12:183–189.

Elkin, R. A. and M. R. Leippe. 1986. "Physiological Arousal, Dissonance, and Attitude Change: Evidence for a Dissonance-Arousal Link and a 'Don't Remind Me' Effect." *Journal of Personality and Social Psychology* 51:55–65.

Emerson, R. M. 1972. "Exchange Theory, Part II: Exchange Relations, Exchange Networks, and Groups as Exchange Systems." In *Sociological Theories in Progress, Vol. 2*, edited by J. Berger, M. Zelditch, and B. Anderson. Boston: Houghton-Mifflin.

Evans, N. J. and P. A. Jarvis. 1980. "Group Cohesion: A Review and Reevaluation." *Small Group Behavior* 11:359–370.

Fazio, R.H. and J. Cooper 1983. "Arousal in the Dissonance Process." In *Social Psychophysiology: A Sourcebook*, edited by J. T. Caccioppo and R. F. Petty. New York: Guilford Press.

Feldman, R. A. 1968. "Interrelationships Among Three Bases of Group Integration." *Sociometry* 31:30–46.

———. 1973. "Power Distribution, Integration and Conformity in Small Groups." *American Journal of Sociology* 79:639–665.

Festinger, L., S. Schachter, and K. Back. 1950. *Social Pressures in Informal Groups*. New York: Harper Bros.

Feynman, R. P., R. B. Leighton, and M. Sands. 1963. *The Feynman Lectures on Physics.* Reading, MA: Addison-Wesley.

Fireman, B. and W. A. Gamson. 1979. "Utilitarian Logic in the Resource Mobilization Perspective." In *The Dynamics of Social Movements,* edited by M. Zald and J. D. McCarthy. Cambridge, MA: Winthrop.

Forsyth, D. 1990. *Group Dynamics,* 2nd ed. Pacific Grove, CA: Brooks/Cole.

Gamson, W. A., B. Fireman, and S. Rytina. 1982. *Encounters With Unjust Authority.* Chicago: Dorsey.

Goodacre, D. M. 1951. "The Use of a Sociometric Test as a Predictor of Combat Unit Effectiveness." *Sociometry* 14:148–152.

Granovetter, M. 1973. "The Strength of Weak Ties." *American Journal of Sociology* 78:1360–80.

———. 1982. "The Strength of Weak Ties: A Network Theory Revisited." In *Social Structure and Network Analysis,* edited by P. Marsden and N. Lin. Beverly Hills, CA: Sage.

Gross, N. and W. E. Martin. 1952. "On Group Cohesiveness." *American Journal of Sociology* 57:546–564.

Hagstrom, W. O. and H. C. Selvin. 1965. "Two Dimensions of Cohesiveness in Small Groups." *Sociometry* 28:30–43.

Hechter, M. 1984. "A Theory of Group Solidarity." In *The Microfoundations of Macrosociology,* edited by M. Hechter. Philadelphia: Temple University Press.

———. 1987. *Principles of Group Solidarity.* Berkeley: University of California Press.

Homans, G. C. 1974. *Social Behavior: Its Elementary Forms,* rev. ed. New York: Harcourt Brace Jovanovich.

Janis, I. L. 1972. *Victims of Groupthink.* Boston: Houghton-Mifflin.

Keller, R. T. 1983. "Predicting Absenteeism from Prior Absenteeism, Attitudinal Factors and Nonattitudinal Factors." *Journal of Applied Psychology* 68:536–540.

Kellerman, H. 1981. *Group Cohesion: Theoretical and Clinical Perspectives.* New York: Grune & Stratton.

Krebs, D. L. 1975. "Empathy and Altruism." *Journal of Personality and Social Psychology* 32:1134–1146.

Lawler, E. J. 1992. "Choice Processes and Affective Attachments to Nested Groups: A Theoretical Analysis." *American Sociological Review* 57:327–39.

Lawler, E. J., C. Ridgeway, and B. Markovsky. 1993. "Structural Social Psychology and the Micro-Macro Problem." *Sociological Theory* 11:268–290.

Lott, A. J., and B. E. Lott. 1965. "Group Cohesiveness as Interpersonal Attraction: A Review of Relationships With Antecedent and Consequent Variables." *Psychological Bulletin* 64:259–309.

Markovsky, B. and M. V. Chaffee. 1995. "Social Identification and Group Solidarity: A Theoretical Extension." In *Advances in Group Processes,* Volume 12, edited by B. Markovsky, K. Heimer, and J. O'Brien. Greenwich, CT: JAI Press.

Newcomb, T. 1961. *The Acquaintance Process.* New York: Holt, Rinehart & Winston.

Nixon, H. L. II. 1979. *The Small Group.* Englewood Cliffs, NJ: Prentice-Hall.

Norris, D. and R. E. Niebuhr. 1980. "Group Variables and Gaming Success." *Simulation and Games* 11:301–312.

Palazzolo, C. S. 1981. *Small Groups: An Introduction.* New York: D. Van Nostrand Co.

Parsons, T. 1951. *The Social System.* New York: Free Press.

Piper, W. E., M. Marrache, R. Lacroix, A. Richardsen, and B. D. Jones. 1983. "Cohesion as a Basic Bond in Groups." *Human Relations* 36:93–108.

Quattrone, G. A. and E. E. Jones. 1980. "The Perception of Variability Within In-groups and Out-groups: Implications for the Law of Small Numbers." *Journal of Personality and Social Psychology* 38:141–152.

Ridgeway, C. 1983. *The Dynamics of Small Groups.* New York: St. Martin's Press.

Scheff, T. J. 1990. *Microsociology: Discourse, Emotion, and Social Structure.* Chicago: University of Chicago Press.

Seashore, S. E. 1954. *Group Cohesiveness in the Industrial Work Group.* Ann Arbor: University of Michigan Press.

Shaw, M. E. 1981. *Group Dynamics,* 3rd ed. New York: McGraw-Hill.

Sherif, M. and C. Sherif. 1953. *Groups in Harmony and Tension: An Introduction to Studies in Intergroup Relations.* New York: Harper and Row.

Silber, L. 1981. "Group Interactions and Group Behavior in Physics." Pp. 56–67 in *Group Cohesion: Theoretical and Clinical Perspectives,* edited by H. Kellerman. New York: Grune & Stratton.

Stokes, J., A. Feuhrer, and L. Childs. 1983. "Group Members Self-Disclosures: Relation to Perceived Cohesion." *Small Group Behavior* 14:63–76.

Sudarshan, E. C. G. 1981. "Cohesion and Physical Structure." Pp. 122–131 in *Group Cohesion: Theoretical and Clinical Perspectives,* edited by H. Kellerman. New York: Grune & Stratton.

Tajfel, H. and J. C. Turner. 1979. "An Integrative Theory of Intergroup Conflict." In *The Social Psychology of Intergroup Relations,* edited by W.G. Austin and S. Worchel. Monterey, CA: Brooks/Cole.

Tallman, I., L. Gray, and R. K. Leik. 1991. "Decisions, Dependency, and Commitment: An Exchange Based Theory of Group Formation." In *Advances in Group Processes,* Volume 8, edited by E. J. Lawler, B. Markovsky, C. Ridgeway, and H. A. Walker. Greenwich, CT: JAI Press.

Turner, J. H. 1992. "The Production and Reproduction of Social Solidarity: A Synthesis of Two Rational Choice Theories." *Journal for the Theory of Social Behavior* 23(3):311–328.

Tversky, A. and D. Kahneman. 1971. "Belief in the Law of Small Numbers." *Psychological Bulletin* 2:105–110.

Wellman, B. 1988. "Structural Analysis: From Method and Metaphor to Theory and Substance." In *Social Structures: A Network Approach,* edited by B. Wellman and S. D. Berkowitz. New York: Cambridge University Press.

Wellman, B. and S. D. Berkowitz, eds. 1988. *Social Structures: A Network Approach.* New York: Cambridge University Press.

Zimbardo, P. 1969. "The Human Choice: Individuation, Reason, and Order Versus Deindividuation, Impulse and Chaos." Pp. 237–307 in *Nebraska Symposium on Motivation,* Volume 17, edited by W. Arnold and D. Levine.

CONFLICT AND AGGRESSION:
AN INDIVIDUAL-GROUP CONTINUUM

Jacob M. Rabbie and Hein F. M. Lodewijkx

ABSTRACT

In this chapter a Behavioral Interaction Model (BIM) is presented that provides an integrative theoretical framework for the social psychological study of intra- and intergroup conflict and aggression. In the first section we discuss the continuity or discontinuity in interpersonal and intergroup behavior, the distinction between social categories and social groups, the antecedents of group formation, the individual-group continuum, the difference between instrumental, and relational interdependence and the levels of analysis at which intergroup relations should be studied. In the second section we examine the relationship between conflict and aggression, problems involved in studying group aggression in the laboratory, and present an outline of the Behavioral Interaction Model. In the third section we discuss empirical evidence for the validity of the individual-group continuum and the differences between individuals and groups in conflict as well as in aggression. In conclusion, the implications of the model and research are briefly indicated.

Advances in Group Processes, Volume 11, pages 139–174.
ISBN: 1-55938-857-9

INTRODUCTION

In this chapter we will review our research on intra- and intergroup relations, particularly on the issue whether groups are more competitive and aggressive than individuals. This research has been guided by a Behavioral Interaction Model (BIM). In this model an attempt is made to integrate a variety of theories in the area of intra- and intergroup relations. We will focus particularly on two seemingly conflicting theoretical approaches to this area: the Interdependence Perspective (IP) and Social Identity Theory (SIT). The IP approach is exemplified by the theory on group dynamics of Lewin (1948, 1952), the Realistic Conflict Theory (RCT) of Sherif (1966), elaborated by LeVine and Campbell (1972), and the interdependence theories of Deutsch (1973, 1982), Kelly and Thibaut (1978), and many other theorists. The SIT approach is represented by the social identity theory of Tajfel and Turner (1979, reprinted in 1986) and the Self-Categorization Theory (SCT) of Turner, Hogg, Oakes, Reicher, and Wetherel (1987), a cognitive elaboration of social identity theory. The SIT approach has been by far the most influential and productive theory on intergroup relations in the last twenty years (Messick and Mackie 1989).

The critical differences between the two perspectives refer to the following questions: whether there is a continuity or discontinuity in interpersonal and intergroup behavior, how groups and social categories should be defined, what the antecedents are of group formation, what types of perceived interdependence between people can be distinguished, and at what levels of analysis intergroup relations should be studied. These key issues will be discussed in some detail.

INTERPERSONAL AND INTERGROUP RELATIONS

Continuity or Discontinuity

Since LeBon (1895) and Sumner (1906) published their work around the turn of the century, there has been a pervasive belief in the social sciences that groups are inherently more competitive and aggressive than individuals. If this pessimistic view is correct, it would imply that it is more difficult to resolve conflicts among groups or nations than among individuals. In contemporary social psychology this pessimistic view about intergroup relations is represented by the work of Tajfel and Turner (1986) and Schopler and Insko (1992).

In their Social Identity approach, Tajfel and Turner (1986) assume that members in social categories or social groups—they use these terms interchangeably—strive to maintain or to achieve superiority over an outgroup on some relevant dimension to achieve self-esteem or a positive social identity: that part of one's self-concept that is derived from the social category or social group one belongs to. In their view, intergroup relations are "essentially competitive." The mere categorization of individuals into groups is sufficient to trigger intergroup discrimination favoring the ingroup. They wrote: ". . . the mere awareness of the presence of an out-group is sufficient to provoke intergroup competitive or discriminatory responses on the part of the in-group" (p. 13).

The main evidence for these assertions is based on allocations in the Minimal Group Paradigm (MGP) developed by Tajfel, Billig, Bundy, and Flament (1971). The minimal groups or social categories in the MGP are considered by Tajfel and Turner as "purely cognitive" (p. 14). Moreover, no functional (perceived) interdependence is said to exist within or between the individual members in these cognitive "groups" or "perceptual categories." In their view, realistic conflicts do not occur in the MGP.

On the basis of the MGP research, Turner (1980) has argued that: "The psychological processes implicated in the transition from interpersonal to intergroup relations tend to shift the social behavior *from baseline fairness to ingroup-outgroup discrimination*. . . the psychological processes distinctly implicated in intergroup behavior do not seem to include motives for equality, fairness or equilibrium" (p. 144, italics added). This statement was made in response to the criticisms of Branthwaite, Doyle, and Lightbown (1979) of the MGP research. They have argued, correctly we believe, that researchers using the standard (MGP) of Tajfel et al. (1971), have given a somewhat distorted picture of their data by overemphasizing the influence of ingroup favoritism: the tendency to allocate more monetary points to anonymous members in the own social category than to members in the other social category but neglecting the equally important strategy of "fairness": to give about as much to individuals in the own category than to individuals in the other social category.

The MGP paradigm builds on earlier research on minimal groups by Rabbie (1964) and Rabbie and Horwitz (1969) as Doise has suggested (1988). In our research with the MGP we have shown that when monetary points have to be allocated, the choice behavior in the standard MGP can be better understood as interpersonal, cooperative intra-group behavior aimed at maximizing individual economic outcomes rather than intergroup behavior designed to maintain or enhance self-esteem as Tajfel and Turner (1986) have claimed (Rabbie, Schot, and Visser 1989; Rabbie and Schot 1989, 1990). On the basis of this research realistic conflict theory cannot be excluded as a viable explanation of the allocations in the MGP (Rabbie 1993b).

The position of Turner (1980) has also been questioned by Van Avermaet and McClintock (1988). They believe that fairness does not necessarily dominate behavior in interpersonal relationships, nor does ingroup bias necessarily dominate decision making in intergroup relations. They conclude on the basis of subjects' allocation behavior in studies of Ng (1984, 1985), and their own research that "fairness plays a major role in determining their allocation decisions in intergroup relations, just as been shown earlier with respect to interpersonal relationships" (p.

424). They propose that one should search for communalities rather than for differences between interpersonal and intergroup behavior.

From an interdependence perspective, Schopler and Insko (1992) have found in a series of ingenious experiments with the Prisoners Dilemma Game (PDG) that intergroup, as compared to interindividual behavior, is more competitive and less cooperative. This so-called "discontinuity effect" is, in their view, partially rooted in an "ethnocentric outgroup schema." They contend that: "The anticipation of interaction with another group instigates learned beliefs, or expectations, that intergroup relations are competitive, unfriendly, deceitful, aggressive and so forth" (p. 129).

Both theories have in common that they refer to intra-individual processes in their attempts to explain intergroup competition and aggression. Tajfel and Turner (1986) believe that an individual motive to "maintain and enhance self-esteem" produces competitive intergroup relations. Schopler and Insko (1992) have claimed that a cognitive, intra-individual "ethnocentric outgroup schema" may account for the greater competitive, deceitful, and aggressive behavior of groups than of individuals.

Consistent with our Behavioral Interaction Model (BIM) we assume, in line with the work of Lewin (1948) and Sherif (1967), that intergroup relations should be studied at different levels of analysis. Intra-individual explanations alone are not sufficient and should be complemented by explanations which take intra- and intergroup processes into account that are often embedded in a sociocultural context. Therefore, we propose that groups are not inherently more competitive and aggressive than individuals but that it will depend on the nature of the interdependence structure *between* the groups and the dominant cognitive, emotional, motivational, and normative orientations of members *within* the groups, whether groups will be more or less competitive and aggressive than socially isolated individuals or will not differ at all from each other (Rabbie and Lodewijkx 1992).

Categories and Groups

In an earlier paper we have argued that a conceptual distinction should be made between social groups and social categories. We wrote: "The view that a social group is a unit capable of acting or being acted upon, of moving or being moved, toward or away from benefits and harms should be distinguished from the view that a group is simply a social category, that is, a collection of individuals who share at least one attribute in common that distinguishes them from others..." (Horwitz and Rabbie 1982, p. 249). This distinction is crucial for the understanding of *out*group favoritism in the minimal intergroup situation (Rabbie, Schot, and Visser 1989), the basic similarity between large scale groupings and face-to-face groups (Rabbie and Horwitz 1988), and the issue of categorization versus attributions in intergroup conflict (Horwitz and Rabbie 1989). In line with this distinction, we view groups as social entities of various sizes and clarity of boundaries, ranging from couples, family units, teams, political parties, to national movements (Horwitz and Rabbie 1989). Following Lewin (1948), we have defined a social group as a "dynamic whole" or social system, ranging from a "compact" unit (which may differ in the degree of intimacy and organization), to a "loose mass" whose members are defined not by their similarities to each other but by a perceived relational and instrumental "interdependence of fate" among its members and with the group as a whole. The members of compact social groups are psychologically bound together by mutually linked interests and by making collective decisions, by which they try to achieve individualistic as well as collective group goals or outcomes (Deutsch 1973; Rabbie and Horwitz 1988; Horwitz and Rabbie 1989; Rabbie 1991a).

Tajfel and Turner (1986) have a completely different conception of a social group. They wrote: "We can conceptualize a group . . . as a collection of individuals, who perceive themselves to be members of the same social category, share some emotional involvement in this common definition of themselves, and achieve some degree of social consensus about the evaluation of their group and of their membership in it" (p. 15). It is clear that they focus on the "individual in a social category" rather than on the social group as a whole as the main unit of analysis. This individualistic, self-centered perspective on intergroup relations is also apparent in the work of other leading identity theorists. For example, Hogg and Abrams (1988) state that the social identity approach "focuses on the *group in the individual*" (p. 3, italics in the original). In line with this reductionist, individualistic approach to intergroup relations they define intergroup behavior as: "the manner in which *individuals* relate to one another as members of different groups" (p. 4 emphasis added).

The main problem of this definition of intergroup relations is of course, that it is very difficult to ascertain, primarily for an external observer but even for the individuals involved, whether in any given social interaction situation, for example in the MGP, individuals "relate to one another" primarily in terms of their membership in a social group or are acting on their own as unaffiliated self-interested individuals in a social category (Rabbie 1993a).

Antecedents of Group Formation

The distinction between categories and groups has important implications for the question how groups are formed. In line with Cartwright (1968, pp. 56–57), we have proposed that the "external designation" of members into a "socially defined category" imposes a "common fate" upon them in the sense that opportunities are given or denied to them simply because of their membership in the socially defined category (Rabbie et al. 1989). We assume that an "interdependence of fate" (Lewin 1948) or a "common predicament" (Sherif 1966) arouses a feeling of group belongingness on the part of the members of the group (Lewin 1948). This feeling of belongingness may motivate some of its members, though not necessarily all, to engage in face-to-face interaction on behalf of the group and its members. Rabbie and Horwitz (1988) have argued that intra-group interaction may result in the emergence of group properties such as cohesiveness, common norms and values, ingroup-outgroup differentiation, shared social identities, common interests, leader-follower relationships and so forth which may culminate in the development of a "compact" or organized group (Sherif 1966). In our earlier work on cognitive and emotional social comparison processes, stimulated by the work of Schachter (1959), we have found that the shared experience of the common fate of being threatened with electric shock, motivates people to seek each other's company (Rabbie 1963; Gerard and Rabbie 1961), a process that marks the beginning of a "compact" social group.

It is interesting to note that in one of his latest publications, Tajfel (1982) has come very close to our own position in his thoughtful comments on the Horwitz and Rabbie (1982) chapter in the book he had edited (Tajfel 1982). He subscribed to our distinction between groups and categories and he gracefully acknowledged that he and his colleagues had employed a "confusing interchangeability of terms" (p. 501). Tajfel also dissociated himself from Turner's (1982) position that the formation of social identity is a precondition for the formation of groups. In agreement with the precedence we gave to perceived interdependence of fate, he wrote that: "interdependence of fate is needed to establish the locus of origin for the development of social identity" (Tajfel 1982, p. 505). Obviously, we share that opinion (Rabbie and Horwitz 1988, pp. 119–120).

The Individual-group Continuum

The conceptual distinction between social categories and social groups has played a central role in our research program on intra- and intergroup relations. This program of research can be characterized as an effort on our part to replicate the results of Sherif's classic intergroup "experiments" under more controlled laboratory conditions.

The classic summer camp studies of Sherif and Sherif (1979) have been a landmark in the social psychology of intergroup relations. However, from a methodological point of view they leave much to be desired (Rabbie 1974, 1982a, 1990). The most important methodological criticism is that intergroup competition and cooperation, have not been manipulated independently from each other but *after* each other. In the absence of appropriate control groups, it is difficult to assess whether the results of Sherif (1966) on ingroup-outgroup differentiation (or the "we-they" demarcation) can be attributed to the conditions listed as "Origins of Ingroup-outgroup Differentiation":

1. The classification of individuals into social categories (Lewin 1948; (cf. "control" condition in Rabbie and Horwitz, 1969); Tajfel et al. 1971).

2. The experience of a positive "interdependence of fate" with the own group and a negative interdependence of fate with the outgroup (Lewin 1948; Rabbie and Horwitz 1969, "experimental" conditions).
3. Intra-group coaction (Rabbie and Lodewijkx 1992; Rabbie, Schot, and Visser 1989) or cooperative intra-group interaction (Rabbie 1991a).
4. Group goal incompatibility, per se (Sherif 1966), for example, the *expectations* to cooperate or to compete with another group (Rabbie and Wilkens 1971; Rabbie and de Brey 1971; Rabbie and Huygen 1974; Rabbie, Benoist, Oosterbaan, and Visser 1974; Rabbie and Visser 1972).
5. The mutual frustration and consequent aggression during the intergroup competition (Lodewijkx 1989; Goldenbeld 1992; Rabbie, Goldenbeld, and Lodewijkx 1992).
6. The realization of group members that one has lost or won from the other group (Rabbie, Lodewijkx, and Broese 1985; Rabbie, Visser, and van Oostrum 1982; Rabbie and Visser 1972).
7. Whether one group lost from another group by legitimate or by illegitimate means (Lodewijkx 1989; Rabbie and Lodewijkx 1987; Rabbie 1989).

The conditions listed here correspond roughly with the different stages of group formation and the types of groups that can be associated with each of these stages. Several defining characteristics of these "groups" are summarized as "Definitions of Groups Ordered Along an Independence-Interdependence Dimension" (adapted from Rabbie 1993b, in press):

1a. From the perspective of an outside observer, a group can be defined as a *social category*: a collection of two or more individuals who have at least one attribute in common that distinguishes them from other individuals (Van Leent 1964; Deutsch 1973; Horwitz and Rabbie 1982).
1b. From an internal perspective, the members of a social category may become a *psychological group* when they perceive themselves (and are perceived by outsiders) as belonging to the same social category, forming a distinctive and bounded social unit, sharing a common identification of themselves (Deutsch 1973; Turner 1982; Tajfel and Turner 1986).
2. When individuals are classified (and labeled) by themselves and others on the basis of a similar sociological characteristic, such as age, sex, occupation, education, color, religion, profession, and so forth, they will be called *sociological categories* (Merton 1968).
3. Coacting individuals in social categories become members of *minimal groups* when they are randomly classified as members of a Blue group relative to another Green group (Rabbie and Horwitz 1969) or are categorized on the basis of their preferences for paintings of Klee and Kandinski (Tajfel et al. 1971).

4. Members of a minimal group become a *social group* when they experience an "interdependence of fate" (Lewin 1948) or a "common predicament" (Sherif 1966), that is become aware of their positive interdependence to one another for achieving some of their goals, interests and outcomes for themselves and for the group as a whole relative to other groups (Rabbie and Lodewijkx 1992).

5. A social group becomes an *interacting group*, when some or all of its members directly interact with each other on their own or on behalf of the whole group in an effort to achieve its interdependent goals, interests and outcomes together (Kelley and Thibaut 1978; Rabbie and Horwitz 1988).

6. Over time, an interacting group may develop into a compact *organized group* which is characterized by a structure of status and role relations among its members, by a set of explicit or implicit norms and values that regulate member interaction, with respect to specific activities, obligations, rights, and entitlements of each of its members and to outsiders (Sherif 1967; Deutsch 1982).

Each of the "groups" listed here form an *individual-group continuum* that can be ordered along an independence-interdependence dimension. At the individual pole of this continuum there are self-centered individuals in social categories who perceive themselves to be minimally interdependent on each other for the attainment of their relational and instrumental outcomes. At the group pole of the continuum there are group-centered individuals in social groups who perceive themselves as maximally interdependent on each other for attaining their relational and instrumental outcomes, not only for themselves but also for the group as a whole. In our view, the group pole of the individual-group continuum is reached when members are willing to subordinate their individual interests to the interests of the social group as a whole at some individual costs to themselves (Rabbie and Schot 1990). As soon as members of a social group interact with each other, group properties emerge that cannot be reduced to the properties of the individual members. Under some extreme circumstances the interests of such an (organized) group may become so important that people are willing to sacrifice their lives for it as has been observed in the platoons of the Israeli and German armies.

We assume that even individuals in social categories, for example in the MGP, perceive a minimal positive sense of interdependence among themselves for at least two reasons (Rabbie and Horwitz 1988). First, the experimenter's interests in dividing them into two "groups" could have suggested to them that differential consequences might befall each group as a whole. Second, Tajfel's subjects knew, as individuals, that "they would receive the amount of money that others had rewarded them" (Tajfel et al. 1971, p. 155). Since people have learned in our society that more weight should be given to desires of ingroup members than of outgroup members (Horwitz and Rabbie 1982), they will tend to give more monetary points to members in the own category than to members in the other category in the normative expectation that these allocations will be reciprocated in the same way (Gouldner 1960). As a consequence of this "normative ingroup schema" a tacit coordination or cooperation will take place with members of the own category leading to a significant "ingroup favoritism" or instrumental cooperation with members in one's own category rather than with members in the other category. Questionnaire data show clear evidence for the operation of this "normative ingroup schema": the learned beliefs and anticipations that more can be expected from "ingroup" members than from "outgroup" members (Rabbie et al. 1989). Obviously such a normative ingroup schema is primarily activated at the individual pole of the individual-group continuum.

As can be seen in the second list, Tajfel and Turner (1986) study "intergroup processes" of individuals in social categories and "psychological groups" that are located near the individual pole of the individual-group continuum. Our interdependence perspective covers the whole range of the continuum from "individuals in social categories" to "members in (organized) social groups." That is the reason why we have used individuals as well as (interacting) groups as decision-making units in order to find out whether groups of different sizes are indeed more competitive and aggressive than individuals. Working with intact "interacting groups" permits the study of the effects of intra-group variables on intergroup relations and vice-versa (Rabbie 1982a, 1982b).

Instrumental and Relational Interdependence

We have argued that group members perceive a greater instrumental and relational interdependence among themselves than members in social categories. What do we mean by these different types of interdependence? In our model we have made a distinction between perceived instrumental-task or utilitarian interdependence and perceived relational or socio-emotional interdependence (Deutsch 1982). In *instrumental cooperation* the main aim is to cooperate with each other to attain economic or other realistic or tangible outcomes. In social or *relational cooperation* the goal is to achieve a harmonious, mutually satisfying relationship with one another as an end in itself rather than an instrument to reach an external goal outside the relationship. The aim is to attain a relational understanding with the other in an attempt to explore the possibility whether the relations may grow in depth or will stay at a more superficial level. The goal of social or *relational competition* is aimed at achieving a positive distinctiveness for oneself, one's own group, organization, or own nation from similar and relevant social units in an effort to achieve prestige, status, recognition, or a positive social identity for oneself, one's group, or one's nation. *Instrumental competition* is designed to obtain more tangible and material outcomes than another party. Similar distinctions can be made between instrumental and relational interdependence with regard to motives and strategies such as instrumental and relational fairness (or equality) and instrumental and relational altruism (Rabbie et al. 1989; Rabbie 1991a).

We have maintained that instrumental or relational orientations of people toward one another can either be viewed as a relatively "stable" personality trait or orientation in which the interaction between genetic and socialization variables may play an important role, or can be considered as a "state" variable that may vary under various circumstances, for example, by being placed in a social category or a social group. For example, we have proposed that males in our culture are socialized to have a rather stable instrumental-power orientation toward other people, while females have learned to have a more relational-moral orientation toward others (Rabbie, Goldenbeld, and Lodewijkx 1992; Eagly 1987). It will depend upon the degree of congruence between these relatively stable orientations and the task-environment in which males and females find themselves, whether they will actually behave, or not, in a more relational or in a more instrumental fashion with each other (Rabbie et al. 1992).

A positive instrumental and relational interdependence structure among individual members in a social group leads to intra-group cooperation and consequent cohesiveness (Deutsch 1982; Rabbie 1991a). A negative instrumental and relational interdependence structure between social groups leads to intergroup competition and mutual hostility (Deutsch 1982; Rabbie 1992). In contrast to the ethnocentrism theory of Sumner (1906), we have shown that intra-group cohesion and outgroup hostility are not invariably correlated between each other as he believed, but should be viewed as independent processes (Rabbie 1982, 1992, 1993a). This means that the mere categorization of individuals into "groups" does not invariably lead to intergroup competitiveness and outgroup discrimination as Tajfel and Turner (1986) have asserted. It will depend on the perceived positive or negative interdependence between them whether groups will cooperate or compete with each other.

Levels of Analysis

Sherif (1966) insisted that groups should be investigated at an "appropriate level of analysis." He wrote: "Our claim is the study of relations between groups and intergroup attitudes of their respective members. We therefore must consider both the properties of the groups themselves and the consequences of membership on individuals. Otherwise, whatever we are studying, we are not studying intergroup problems" (p. 62). Lewin (1948), Deutsch (1973), Doise (1986), and Rabbie (1982a, 1992) have also stressed the importance of a multilevel analysis of intergroup relations.

In our research we have always followed a multilevel approach to the study of intergroup relations. At the *intra-individual* level of analysis we have found that individual differences in value orientations, for example, the propensity of individuals to make cooperative, competitive, fair, or altruistic choices, strongly affects the allocations of individuals in the MGP (Lodewijkx 1992: Mlicki 1993). Very similar findings have been obtained by Platow, McClintock and Liebrand (1990) and Chin and McClintock (1993). These results indicate, once again, that the allocations in the standard MGP should be interpreted more as interpersonal rather than intergroup behavior. In our opinion they are more closely located near the individual rather than near the group pole of our continuum.

In our PDG research on conflict between individuals and groups, we asked the subjects prior to the actual exchange of choices whether they planned to make a cooperative C-choice or non-cooperative D-choice in the game. They were classified on the basis of their prior intentions to make C- or D-choices and what expectations they had about the C- or D-choices of the other player. It turned out that the CC players, whether they had to act as individuals or as group members, behaved more as "cooperators" and the DD players more as "competitors." Consistent with the triangle hypothesis of Kelley and Stahelski (1970), cooperators were more likely to follow a more contingent, cooperative tit-for-tat (TFT) strategy in the PDG, while competitors made more noncooperative D-choices, regardless the behavior of the other party (Rabbie, Visser, and van Oostrom 1982). In a TFT strategy, an actor (an individual or a group), starts with a cooperative choice and thereafter does whatever the other player did on the previous move (Axelrod 1984). A TFT strategy is considered to be the most effective procedure to maximize one's own and joint outcomes in the PDG (Wilson 1971). It appears that individuals as well as groups have cooperative or competitive dispositions. Thus, groups are not invariably more competitive than individuals as Tajfel and Turner (1986) believe. There are clear group differences in this respect.

In our aggression studies we have found that a strong instrumental orientation, of both men and women, measured prior to the experimental manipulations, tends to enhance their aggression, in individual as well as intergroup relations. A strong relational orientation tends to reduce or inhibit aggression (Lodewijkx 1989; Goldenbeld 1992; Rabbie et al. 1992).

At an *intra-group level of analysis*, we have shown that leaders who felt threatened in their leadership position are more likely than stable, nonthreatened leaders to opt for intergroup competition rather than for intergroup cooperation in an effort to unify the group behind themselves and to maintain their leadership position. This also depends on the relative power relationships between the groups involved. The stronger the negotiation position of the group and the more likely it appeared that the intergroup conflict could be resolved to the own group's advantage, the more the threatened leaders choose for intergroup competition rather than for intergroup cooperation in their negotiations with the other group, in an effort to maintain their leadership position (Bekkers 1977; Rabbie and Bekkers 1976).

The insistence that intergroup relationships should be studied at different levels of analysis requires a multi-method, convergent approach in which laboratory research, case studies, and survey methods have to be used in different sociocultural settings. That is the reason why we have studied intergroup relations between members of different ethnic groups (With and Rabbie 1985); between different national groups (Mlicki 1993); between departments in educational institutions (Rabbie, Visser, and Vernooy 1976); between new and old members in student associations (Lodewijkx and Akkersdijk, 1993); in labor-management relations (Bekkers 1977), and field experiments on the relations between the riot police and demonstrators (Kroon 1992; Kroon, van Kreveld, and Rabbie 1991, 1992).

However, we do have some preference for laboratory research. When we create ad hoc groups in the laboratory it is easier to assess the causal effects of the experimental manipulations on such phenomena as ingroup-outgroup differentiation than using members of "natural groups" or members of "sociological categories" (Merton 1968), which have been socialized to feel and react in different ways to members of the own and the other group. But even in laboratory experiments, the influence of sociocultural factors cannot be excluded. That is the reason why we have been so interested in studying sex differences in conflict and aggression that cannot be dissociated from the sociocultural socialization processes that influence the cognitive, emotional, motivational, and normative orientations and behavior of males and females who grew up in individualistic and collectivistic cultures (Eagly 1987; Triandis 1989; Markus and Kitayama 1991).

CONFLICT AND AGGRESSION

Social conflict between individuals, groups, organizations, and other social systems are defined in terms of perceived incompatible priorities in interests, goals, preferences, and values between interdependent parties that cannot be realized simultaneously (Pruitt and Rubin 1986). These interests or priorities may involve instrumental, tangible, or realistic outcomes or may refer to relational or symbolic outcomes such as status, prestige, or recognition (Lewin 1948; Sherif 1966; Rabbie and Schot, 1989).

Aggression is viewed as a form of social conflict between interdependent parties in which one of the parties, the actor, intentionally hurts or harms another victim psychologically or physically (Baron 1977). Instrumental aggression is considered as a type of conflict behavior between various parties who intentionally inflict harm on the other party, or threaten to do so, in an effort to settle the conflict to their own advantage. Hostile, angry, or reactive aggression originates in feelings of anger, indignation, or revenge that is designed to harm another party on purpose; a goal that can only be achieved by aggression (Buss 1961; Feshbach 1964). An offensive, unprovoked aggressive attack of another party is often considered as unjustified and illegitimate. A defensive reaction to such an offensive attack is seen as a form of defensive aggression or punishment that is viewed as a morally acceptable and legitimate reaction.

In our research we assume that angry or reactive aggression occurs as a reaction to the norm violation of another party, an individual, or a group. This notion was suggested by the findings of Sherif (1966). The groups of boys in his summer camp studies reacted with anger and aggression to the outgroup when Sherif had made one group believe that the other group had treated them in an unfair and callous way. This finding indicates that intergroup competition and consequent frustration does not necessarily lead to open aggression but that at least one party has to make the attribution that another party takes advantage of him or her in an unfair and illegitimate way (Ferguson and Rule 1983). These unfavorable attributions may lead to the escalation of mutual conflict and aggression between the parties (Horwitz and Rabbie 1989).

This assumption is also a central tenet in the frustration-aggression (F-A) theory of Brown (1986) and Brown and Herrnstein (1975). Their theory can be considered as a revision of the classic F-A theory developed by Dollard, Doob, Miller, Mowrer, and Sears (1939). Brown and Herrnstein (1975) have argued that frustration can be conceptualized as an "illegitimate disappointment of legitimate expectations" (1975, p. 274). In their view, frustration or disappointment occurs when people, individuals, or groups, have legitimate expectations about their outcomes that were created at an earlier time and that happen to fall short of their actual outcomes at a later time. In this sense, the F-A theory has always had "a between-times or historical aspect." They also contend that not any frustration or disappointment leads to aggression but that the frustration—the "blocking of a goal response" by another party—should be seen as an unjustified and illegitimate response of the other party. In our own terminology, angry aggression is aroused as a reaction to social injustice, in a response to an unprovoked norm violation by another party (Rabbie and Lodewijkx 1991).

In a later publication Brown (1986) has argued that ethnocentrism, that is thinking more highly of ingroups than of outgroups, is not sufficient to bring about hostile actions between groups. It is also necessary that groups are similar enough so that they can compare their outcomes, for example, the distribution of rewards and costs between each other, which may lead to the conclusion that these outcomes are unfair and inequitable. In his view, "*Unfair* or unjust disadvantage or frustration is the sovereign cause of anger and aggression. Disadvantage or frustration will not alone do it; perceived justice is critical" (Brown 1986, p. 534). We agree with this view, but believe that procedural injustice as well as distributive injustice is involved in the arousal of reactive or angry aggression (Rabbie and Lodewijkx 1991; Syroit 1991).

The comparison of outcomes are not always based on the past history of the same group or individual. They can also be made with reference to a relevant comparison group or individual as most relative deprivation theorists have assumed (e.g., Sherif and Sherif 1964; Merton 1968; Gurr 1970; Crosby 1976). In our experiments, individuals and groups were not only frustrated in their legitimate expectations by the illegitimate norm violation of another individual or group, but they also suffered a sense of relative deprivation, when they compared their own outcomes with those of the other party. The experience of relative deprivation would be greater the more they felt that they had lost from the other party. Since aggression may be one of the responses to relative deprivation, particularly if the other party cannot retaliate in kind, it is to be expected that the greater the loss or frustration the subjects experience by the illegitimate actions of the other party, the greater the angry aggression would be (Rabbie and Lodewijkx 1987). Testing these hypotheses requires a discussion of the methods we have used to verify the norm violation theory of Brown (1986; Rabbie and Lodewijkx 1991).

Group Aggression in the Laboratory

Group aggression is one of the most urgent problems of our time. Against this background, it is surprising that most research in experimental social psychology has dealt with individual rather than with group aggression. The focus is on intra-individual processes which may account for aggressive behavior. Very little attention is paid to the nature of social interaction processes within and between the groups involved which may lead to intergroup aggression.

Systematic comparisons between individual and group aggression are very rare indeed (Jaffe and Yinon 1983). One of the reasons for this neglect might be that it is very difficult to create group aggression in the laboratory. For ethical and practical reasons it is impossible to produce the kind of group aggression and violence we read about in our daily newspapers. We have to work with much milder forms of physical and psychological aggression in the laboratory. Moreover, it is also much more expensive to use groups rather than individuals as the main units of analysis. There are also theoretical and methodological difficulties in studying group aggression in the laboratory.

With regard to the theoretical difficulties it should be noted that research on intergroup relations on topics like stereotypes, prejudice, and discrimination, have been dominated by an individualistic, cognitive information processing approach that studies the way in which information about groups or social categories and their members are mentally represented in the individual (Messick and Mackie 1989). The emphasis of this approach is how these mental representations are encoded, stored in memory and retrieved and in what way they affect the judgment of the individuals about groups and their members. How these judgments affect actual intergroup behavior is seldom investigated. The approach is almost exclusively cognitive, there is little or no attention to the motivational, emotional, and normative orientations that may determine interpersonal and intergroup behavior and it tends to neglect the historical and sociocultural context in which group aggression occurs (Stroebe and Insko 1989; Rabbie 1992).

There are also methodological problems. Most of the studies on individual aggression for the last twenty years have used the "aggression machine" in the context of the "teacher-learning paradigm" developed by Buss (1961) and Milgram (1974). In this research tradition, male and female college students take the role of a teacher who has to deliver electric shocks or other aversive stimuli, such as irritating noise, to punish the "learner" for apparent errors. To the subjects the learner appears to be a subject like themselves. In fact, he or she is a confederate of the experimenter who makes standard errors in a learning task. The intensity, duration, and frequency of the noxious stimuli administered to the "learner" are used as indications of physical aggression.

As we have argued elsewhere, it is debatable whether this paradigm studies aggression at all (Rabbie and Lodewijkx 1987). It has been shown that subjects consider their shock behavior not as an aggressive act but as altruistic behavior, helping the learner to accomplish his or her learning task (Baron and Eggleston 1972). Thus, it is not clear whether the "aggression" was intended to harm or hurt the other, a crucial element in the definition of aggression (Baron 1977). The real purpose of the experiment in the teacher-learner paradigm is carefully concealed from the subjects. Unlike the subjects in our experiments, they are *not* told that they can deliver painful white noise to the other as a way to express disapproval of the other's behavior. Most importantly, the aggressive behavior in this traditional paradigm is not part and parcel of a continuous interaction process between the actors, in which aggression is only one of the responses they can make to settle the conflict among themselves. In this paradigm there is no way to study the history of the conflict and the impact that it may have on the aggressive behavior of individuals and groups over time (Rabbie et al. 1992).

In our research we have used experimental games, such as the Prisoner's Dilemma Game (PDG) or the Power Allocation Game (PAG) (Rabbie and Lodewijkx 1991), in an attempt to correct these deficiencies. These games permit the comparison of the conflict and aggressive behavior of individuals and groups with each other.

Before describing a few illustrative experiments on conflict and aggression, we have to give a brief outline of our Behavioral Interaction Model that has guided our research on individual and group aggression in recent years.

The Behavioral Interaction Model

In view of space limitations we will give only a brief outline of our model. For a more extended treatment of the model the reader is referred to analyses of armed conflicts between nations (Rabbie 1987); aggressive reactions to social injustice (Rabbie and Lodewijkx 1991); intra-group cooperation (Rabbie 1991a); terrorism (Rabbie 1991b) and the origins of intra- and intergroup conflict (Rabbie 1993a).

In our Behavioral Interaction Model, depicted in Figure 1, it is assumed, consistent with the interactionist position of Lewin (1952) that behavior, including the cooperative or noncooperative behavior of subjects in the PDG or the aggressive reaction to the norm violation of another party, is a function of the external environment and the cognitive, emotional, motivational, and normative orientations that are in part elicited by the external task environment and in part acquired by individuals and groups in the course of their development (see Figure 1, upper half). The main function of these psychological orientations is to reduce the

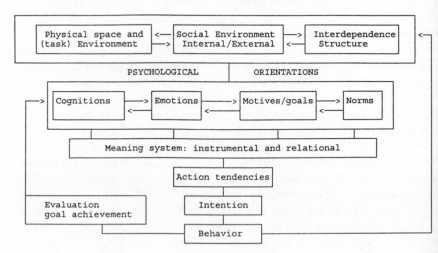

Figure 1. The behavioral interaction model

uncertainty in the external environment to such a level that it enables individuals, groups or organizations to cope effectively with the environment in an effort to achieve desirable, and to avoid undesirable outcomes. The external environment consists of three components: (1) a physical (task) environment; (2) an internal and an external social environment: such as the behavior of other people within and external to the group or social system; and (3) a positive or a negative interdependence structure between the parties that may help or hinder the parties in reaching their goals. The goal interdependence between the parties may be loosely or tightly coupled and may be symmetrical or asymmetrical with regard to the power relations between the parties. In asymmetric power relationships, party A depends more on party B for reaching his or her outcomes than vice-versa.

These different psychological orientations produce a meaning structure or interpretative system about the situation which in turn generates various action tendencies in the actor or party: an individual or a group. Although many types of meaning systems may exist, we have focused our attention on instrumental and relational orientations that combine with the different cognitive, emotional, motivational, and normative orientations that have been distinguished in the literature (Deutsch 1982). In this context we have made a distinction between instrumental and relational cooperation, competition, fairness, and altruism.

Consistent with various value-expectancy models (e.g., Lewin 1952; Porter and Lawler 1968; Ajzen and Fishbein 1980), it is assumed that among competing action tendencies and available strategies, those actions and alternatives will be chosen that promise, with a high probability of success, to attain the most valued goals or profitable outcomes, whereby the gains seem to exceed the costs of achieving them. Thus, it can be expected that people will choose to cooperate rather than compete with each other when mutual instrumental cooperation seems to be more profitable to them than mutual (instrumental) competition in an effort to maximize their individual, as well as their collective outcomes (Pruitt and Kimmel 1977). This view implies that individuals, and groups in particular, come more often to their decisions, in the instrumental rational environments of experimental games, by a careful and thoughtful consideration of the true merits of the alternatives, that is, are more likely to follow the "central route" of information processing, than use a more "peripheral route" to arrive at their decisions (Petty and Cacioppo 1986).

The action tendencies, depicted in Figure 1 (lower half), result in an intention: the commitment to perform or not to perform a certain action. According to the theory of reasoned action (Ajzen and Fishbein 1980) or planned behavior (Azjen 1988), the intention is the best predictor for the behavior. The action or the behavior may lead to outcomes or a present state that has to be evaluated against a desired future state: the goal or standard the party desires to achieve. When no discrepancy is observed between the present and the desired state, the action sequence is terminated or "exited." When a discrepancy is observed between the present and desired state, the party has to revise or reconsider his or her psychological orientations that led to the behavior which was not successful in reaching the earlier standards or goals the actor tried to realize. The mismatch between the behavioral outcomes and goals may induce a change in the psychological orientations symbolized by an arrow from the "evaluation of goal achievement-box" to the situational meaning system in Figure 1 (to the left). This may initiate a new action sequence that will be terminated until the goal is achieved or when different or more reachable goals are substituted for the original objectives.

The direct arrow at the right hand side of the Figure, from the behavior to the external environment at the top of Figure 1, reflects the notion that acting on the external environment may have the effect of changing it, leading to different psychological orientations and meaning structures of the situation, which in turn may induce different action tendencies, intentions, and behaviors, initiating a new cycle in the action sequence until the standards or goals, either the original ones or the modified goals, are achieved. Just like other cybernetic action models, for example, the control model of Carver and Scheier (1981) or the TOTE-concept of Miller, Galanter, and Pribram (1960), our model can be considered as a self-regulating, negative feedback system. The model is particularly useful in tracing the development of the conflict over time, for example by examining the changes in psychological orientations and behavior in the different trials of a PDG (Rabbie et al. 1982; Visser 1993).

EMPIRICAL STUDIES ON CONFLICT AND AGGRESSION

Conceptually and methodologically we have viewed aggression as an outgrowth or consequence of the preceding social conflict between the parties. That means that current aggressive outbursts cannot be understood without taking the history of the conflict into account. In our first series of studies on conflict with the PDG we were mainly interested in the question whether groups are always more competitive than individuals. In the second series of experiments we used the PDG to study aggression among individuals and groups as well.

In view of space limitations and to give the reader an impression of the development of our arguments, we will discuss a few experiments in some depth that may illustrate the possible usefulness of our model rather than trying to summarize all the experiments that have been conducted on these issues. That also means that this discussion will be restricted to research with the PDG and MGP. The PAG studies are reviewed elsewhere (e.g. Rabbie and Lodewijkx 1991; Rabbie et al. 1992; Goldenbeld 1992).

In the first study to be reported, experimental evidence is provided for the validity of our individual-group continuum. In the next section we will focus on the question whether groups are more competitive than individuals. In the third section the issue will be whether groups react with more angry aggression to the norm violation of the other party than individuals.

Empirical Evidence for the Individual-Group Continuum

In an effort to test the validity of the continuum hypothesis we have used the Minimal Group Paradigm (MGP) of Tajfel et al. (1971) in a recent experiment of Rabbie and Schot (1989). The physical task environment of the standard MGP has a highly instrumental, strategic, and impersonal character. Individuals participating in the MGP are socially isolated. They are seated in separate cubicles and are asked to allocate (monetary) points to anonymous individuals in their own social category and to anonymous individuals in another social category. They do not receive immediate feedback about the effects of their allocation behavior, so they do not know what kind of impact their behavior has on other people. We assume that in this highly impersonal, individualistic, and socially impoverished task environment, people are mainly motivated to maximize their own individualistic outcomes. They use the social categories provided by the experimenter as a vehicle or tool to reach that goal.

The degree of perceived interdependence along the individual-group continuum in this MGP experiment was varied in the following way. In the minimal interdependence or *Individual condition*, the subjects were urged to maximize their economic self-interests. In the group-goal interdependence condition or *Group condition* the subjects were instructed to maximize their group interests as well as they could. In the *Control condition* the standard instructions of Tajfel et al. (1971)

were given which do not refer to the maximization of individual or group interests. The "groups" in the *Individual* and *Control condition* can be better labeled as "perceptual" or "social categories" rather than as social groups as we have defined them. Therefore, both conditions can be considered as category conditions that are located near the individual pole of the individual-group continuum. We assume that subjects in the standard MGP are mainly motivated to maximize their individual outcomes, whether they are explicitly told to do so, as this occurred in the individual condition, or whether they simply asked to participate in the standard MGP. Thus no difference in allocation behavior between the Control and the Individual conditions was expected, but allocations in both category conditions would be different from the allocations in the Group (goal) condition.

The interdependence structure of the MGP has more of a zero-sum, or win-lose character than the mixed-motive structure of the PDG. Therefore, it was expected that subjects would experience a positive interdependence of fate with their fellow ingroup members but a negative interdependence of fate with members of the other outgroup members. Consequently, we predicted that there would be more instrumental *intra*-group cooperation and more *inter*group competition and less instrumental intergroup fairness in the group than in the two category conditions.

These hypotheses received strong support. As expected, no difference in allocation behavior was found between the individual and the control condition. These results suggest again, contrary to the claims by Turner (1982) and Tajfel and Turner (1986), that in the standard MGP of Tajfel et al. (1971), we are dealing more with instrumental individual behavior rather than with relational intergroup behavior designed to maintain or enhance one's self-esteem.

Moreover, in support of our expectations, more intra-group cooperation and more intergroup competition and less (instrumental) fairness occurred in the group—than in the category conditions. Subjects in the group condition acted in the interests of the group as a whole at some individual costs to themselves, suggesting that they can be located as behaving near the group pole of the individual-group continuum. Questionnaire results indicated that subjects in the category condition perceived a greater sense of "independence than of interdependence", and were more "acting as individuals than as group members" than subjects did in the group condition. Group members also reported a greater sense of belongingness and ingroup identification in the group than in the category conditions. Consistent with our hypotheses, subjects reported that they were more motivated "to be better than the other group" and were "trying to enhance their self-esteem" more in the group than in the category conditions.

These results challenge, once more, the assertions by Tajfel and Turner (1986) that in the MGP there is no ". . . conflict of interests . . . nor is there any rational link between economic self interests and the strategy of ingroup favoritism" (p. 14). In this and in many other experiments we have shown that there are, in fact "realistic conflicts" between individuals in social categories and that these individuals do construct "a rational link" between their economic self-interests and their allocation behavior and are observed to act on these perceptions (Rabbie et al. 1989; Rabbie and Schot 1989, 1990; Schot 1992; Lodewijkx, Mlicki, Schot, and Rabbie 1991; Mlicki 1993).

The finding that the goal interdependence among members in social groups increases their feelings of belongingness and ingroup identification, as compared to individuals in social categories, has been replicated several times in studies using the MGP (Mlicki 1993; Lodewijkx et al. 1991) or employing the PDG (Rabbie and Lodewijkx 1992; Lodewijkx 1989). Thus, there can be little doubt that the individual-group continuum is a valid and reliable phenomenon.

Individuals Versus Groups in Cooperation and Competition

As we have indicated earlier there is a common belief that groups are more competitive in their motivational orientations and behavior than individuals. Research on intergroup relations in experimental social psychology seem to support this point of view. In an early review of this research it is concluded that ". . . there is some consistent evidence that social groups seem to be more competitive and perceive their interests more competitively than individuals under the same functional conditions" (Turner 1981, p. 7). In a recent review, Schopler and Insko (1992) have concluded, based on their research with the PDG, that groups are invariably more competitive and less cooperative than individuals. They propose that there is a basic discontinuity between interpersonal and intergroup behavior in this respect.

We have always been doubtful about the generality of these propositions. According to our model, it will depend on the nature of the interdependence structure between the parties and the kind of attributions that are made about the expected or actual behavior of the other parties whether groups will be more or less competitive than individuals or will not differ at all from each other (Rabbie et al. 1982; Rabbie 1987).

In their goal expectation theory, Pruitt and Kimmel (1977) have proposed that individuals and groups in a PDG will strive for the long range goal of mutual cooperation—by making mutual C-choices in the game—provided that the other party can be expected or trusted to reciprocate these cooperative initiatives. If there are reasons to believe that the other will make a competitive or defensive D-choice, the own party will make a defensive D-choice as well, trying to minimize a maximum possible loss (McClintock 1972). Pruitt and Kimmel (1977) have suggested that a greater insight in the interdependence structure of the PDG will lead to more mutual cooperation. They assume that intra-group discussions or the filling out of questionnaires about their choices will make people more aware of the reward—or interdependence structure of the PDG, leading to the insight that in a PDG it is more profitable for both parties to strive for the long range goal of mutual cooperation than to defect or compete with each other. This means that groups are more likely to use a cooperative tit-for-tat (TFT) strategy than individuals, that is, they tend to make a first cooperative choice and then they will match the choice the other has made on the previous trial in an effort to influence and shape the other's behavior in a more cooperative direction (Rabbie 1991a). It has been shown that this contingent, cooperative TFT strategy is the most effective way to maximize one's outcome by mutual cooperation (Wilson 1971; Pruitt and Kimmel 1977; Axelrod 1984). This notion would imply that groups will show a greater goal-rationality or "Zweckrationalität" (Weber 1921) than individuals. They are more likely than individuals to react with defensive competitive orientations and behavior toward a competitive or defensive opponent but will react more cooperatively in their orientations and behavior to a cooperative opponent.

This hypothesis was tested in a series of experiments reported by Rabbie et al. (1982). In one experiment male subjects, individual, and dyads, played a five trial PDG in which the parties had to make their choices simultaneously and independently of each other. The other programmed opponent played the following pattern of choices: C,D,D,C,C. Individuals made decisions on their own while dyads made collective decisions in the game. No significant differences were obtained in the first three trials. However, after having received two competitive and exploitative choices in a row after an initial C-choice, dyads made significantly more competitive or defensive D-choices on the two last trials of the game. Thus, consistent with our hypothesis, groups or dyads are more likely to make more competitive choices than individuals, if they are faced with a competitive and exploitative opponent.

In a second experiment a three-member group, or triad, was added and this time we worked with males as well as with females to find out whether there would be any sex differences in individual and group behavior. Earlier research had shown, contrary to the usual sex stereotypes (Colman 1982), that individual males make more cooperative choices in the PDG than individual females (Rapaport and Chammah 1965). The game was extended to a seven trial game in which the programmed opponent played C,D,D,C,C,C,C choices. This extension allowed us to examine how long the defensive behavior of the parties would persist after the fourth critical trial at which point the subjects would have received two consecutive D-choices in a row from their opponents.

The results show that: "all (100%) of the male and female dyads chose a D on the fourth and critical trial, while 44 percent of the individuals and 56 percent of the triads chose this course of action ($p < .002$). As expected, dyads chose more D's than individuals, but surprisingly these data also indicate that they chose more D's than triads" (Rabbie et al. 1982, p. 325). Questionnaire data indicated that intragroup interaction, especially among dyads, creates a greater awareness of the risky aspects of the PDG than in individuals or triads.

Consistent with earlier work of Rapaport and Chammah (1965), males, whether they acted as individuals or as groups, behaved more cooperatively than females, particularly when they were faced with a cooperative choice of their opponent. This result is consistent with the work of Wilson (1971) and the interdependence theory of Kelley and Thibaut (1978) who have pointed out that males are likely to use a more contingent, TFT strategy than females. Thus it seems that the more instrumental groups and males have a greater insight in the interdependence structure of the PDG (Pruitt and Kimmel 1977) and are therefore more likely to follow an instrumental TFT strategy than the more relational individuals and females (Rabbie 1991a; Rabbie et al. 1992).

In a recent experiment (Lodewijkx and Rabbie 1992), we tested hypotheses which seem at variance with the discontinuity hypothesis of Schopler and Insko (1992) which holds that groups behave invariably more competitively and less cooperatively than individuals. In line with our model we have a different view. If groups, and particularly dyads, are likely to follow a TFT strategy (Pruitt and Kimmel 1977; Rabbie et al. 1992), it is to be expected that groups will tend to follow a more responsive, instrumental *cooperative* TFT strategy than individuals, provided that the other party can be expected to behave cooperatively as well. According to the conflict avoidance hypothesis of Rabbie et al. (1982), members of dyads will choose that alternative which minimizes the internal conflict between them. Such internal conflicts may arise if the dyad is confronted with a threatening competitive opponent, but it will be absent when they face a benign, cooperative opponent as was the case in the present experiment. In the studies of Rabbie et al. (1982), we have shown that dyads react more defensively and competitively than triads and individuals to the exploitative behavior of the opponent. Through this behavior members in the dyad minimize losses and maximize gains and avoid internal conflicts among themselves. If the other party appears to behave cooperatively, members of dyads will tend to reciprocate the cooperative behavior, because in this way joint profits are maximized and internal disagreements will be prevented. Thus, according to our model, actual intergroup competition or—cooperation is viewed as a joint function of factors operating in the internal environment of a group and attributions that are made about the external social environment: the way the opponent is expected to behave. When the other party seems to follow a consistent cooperative strategy it is expected that dyads will be more instrumentally cooperative in their choice behavior and orientations than individuals. When the other party does not reciprocate the opposite can be expected.

These expectations are at variance with the discontinuity hypothesis of Schopler and Insko (1992). They believe that the anticipation of interaction with an outgroup always triggers a competitive, ethnocentric outgroup schema. Following the discontinuity hypothesis, one would assume that groups faced with another group, would expect outgroups to behave more competitively than individuals would vis-à-vis an individual opponent even *prior* to the actual exchanges of choices between the parties. To make sure that the other party would be perceived as acting cooperatively, in the first sequential trial of the three trial PDG, the programmed opponent made a cooperative C-choice to which the subjects could respond. In the second trial a simultaneous play was used in which, again, the other programmed party made a cooperative C-choice. In the third and last sequential trial the subjects had to make the first choice to which the subjects could react. Prior to making the third choice the subjects received a message of the other programmed party, via a TV screen, that they would make a cooperative C-choice again. The reactions of the subjects to the first trial are the most critical for testing the discontinuity hypothesis of Schopler and Insko (1992). If group members are indeed inclined to use a cognitive, ethnocentric outgroup schema, it is to be expected that they are more motivated to exploit the cooperative first choice of the opponent by making more competitive D-choices than individuals would do.

In contrast to the discontinuity hypothesis of Schopler and Insko, groups (triads *and* dyads), made, in fact, significant more C-choices at the first trial in response to the first C-choice of the other party than individuals. In support of our second hypothesis, it is found that these differences can be more attributed to the difference in C-choices between individuals and dyads ($p < .01$), and not so much to the difference between individuals and triads ($p < .17$).

The discontinuity hypothesis was tested directly, by asking individuals, dyads, and triads, prior to any group interaction or exchange of choices, whether they intended to make a C- or a D-choice and what choice they expected the other individual, dyad, or triad to do. Contrary to the discontinuity hypothesis it we found that groups, particularly dyads, intended to behave more cooperatively than individuals prior to any social interaction between them. Moreover, dyads (46%) and triads (36%) indicted more often than individuals (26%) that they planned to make a cooperative C-choice and expected a C-choice in return (CC). Thus, in accordance with our model, groups are more inclined than individuals to use a cooperative TFT strategy in a PDG, contrary to the discontinuity hypothesis of Schopler and Insko (1992).

In the Lodewijkx and Rabbie paper (1992) we discuss at length the reasons for the apparent differences between the results of Schopler and Insko (1992) and our own data. In our view, there are crucial differences in the experimental procedures used by the two research groups which may account for the divergences in the results. Since these two groups of researchers share the same theoretical interdependence perspective, it is our contention that theoretically our views are essentially compatible with each other, but more research is needed to substantiate that claim.

Aggression Between Individuals and Groups

Consistent with the revised frustration-aggression theory of Brown (1986), we have argued that it is the (illegitimate) norm-violation of another party rather than competition per se that arouses angry aggression in individuals and groups. According to our enhancement of group polarization hypothesis, derived from our model (Rabbie and Visser 1972; Rabbie and Lodewijkx 1985), it is expected that intra-group interaction enhances the cognitive, emotional, motivational, and normative orientations that are already present in single individuals prior to their participation in the group discussion.

In two experiments we have found some evidence for his hypothesis (Rabbie and Horwitz 1982). In these studies a two-trial and two-party PDG was used. In the first experiment, male individuals and dyads had to make an individual or a collective decision in the PDG. Each individual was paired with another single male, while a dyad was always pitted against another dyad. The other party could be seen on a TV screen in front of the subjects. Unknown to them, the other (programmed) party was prerecorded on video tape. At the first (sequential) trial, the programmed party always made a cooperative choice to which the subjects could respond by making a cooperative C-choice or a defecting D-choice. Prior to the second trial, during which both parties made a simultaneous choice, the parties could send messages to each other. The subjects received a message of the programmed party promising that he would again make a cooperative choice on the second trial and requesting the other party to do the same. The subjects could respond with a message of their own before making their own choice on a payoff matrix whose values were increased by about half. In the *norm-violation condition* the other party broke his promise by making a competitive rather than a cooperative choice. In the *norm-adherence condition*, the opponent kept to his promise and made a cooperative choice. After the choices were exchanged, the subjects could react to the behavior of the other party by delivering (painful) white noise to him or them. The noise had been described to them as a way to communicate with the other party. In this way they could indicate "the extent of their disagreement." The other party had no white noise apparatus at their disposal and could not retaliate to the actions of the subjects. The subjects knew what they were doing to the other party since they had heard two samples of noises described as very soft and pleasant or very irritating and almost unbearable.

It was expected that a breach of promise to make a cooperative choice would evoke more angry aggression than when the other dyad had kept to his promise by making a cooperative choice. As expected individuals and groups reported more feelings of anger, tension, and uncertainty when the other party broke the promise than when he did not. Moreover, subjects in the *norm-violation* condition considered the other party more hostile, less fair, morally wrong, more cowardly, too interested in earning money, and more dishonest than subjects in the *norm-adherence* condition. All these differences were significant beyond the $p < .001$ level of statistical significance, using an analysis of variance.

Consistent with our hypothesis, more painful white noise was delivered to the other in the norm-violation condition than in the norm-adherence condition. As expected, dyads reacted with more angry aggression to the deceptive behavior of the other party than individuals did.

The *norm-violation* condition differed from the norm-adherence condition in at least two ways. The *norm-violator* broke his promise and played competitively, while the non-violator kept his promise and played cooperatively. It is therefore impossible to ascertain whether the difference in aggression obtained between dyads and individuals reflect differences in their reaction to the competition rather than to the norm violation. In a second experiment an attempt was made to disentangle the effects of the two components by varying the breach of promise but keeping the competitive behavior of the opponent constant. On the first trial the programmed party again made a cooperative choice to which the subjects could respond. Prior to the second trial, subjects in the *breach-of-promise* condition, but not in a *no-promise* condition, received a promise that their opponents would play cooperatively. Subjects than made a first move on that trial. In both conditions, the opponents always responded with a competitive D-choice. Subjects could than administer the white noise which, at this time, was described as enabling them to express "the degree of approval or disapproval" of the behavior of the other party.

Both individuals and dyads delivered white noise of higher intensities in the *breach-of-promise* than in the *no-promise* condition. Again the individuals and dyads had a more negative view of the other in the promise than in the no-promise condition. As expected dyads reacted more aggressively to the breach-of-promise than individuals, but not differently from individuals where the opponent had made no promise at all. These results indicate that the greater aggression by dyads rather than individuals is due to the difference in how they responded to the norm violation and not to the difference in how they reacted to the competitive behavior of the opponents. In accordance with our enhancement hypothesis, intra-group interaction induced more aggression. In other experiments, very similar findings have been obtained (e.g., Rabbie and Lodewijkx 1987, 1991; Rabbie, Lodewijkx, and Broese 1985; Lodewijkx 1989; Rabbie et al. 1992; Goldenbeld 1992; Kroon, van Kreveld, and Rabbie 1992a,b).

In the previous experiments, it was shown that the norm violation, not the competitive behavior of the other programmed opponent aroused more aggression in groups (Rabbie and Horwitz 1982). One possible explanation of this finding could be that groups attach greater weight to their collective interests than individuals do to their own selfish interests (Horwitz and Rabbie 1982). This assumption is based on the notions of Simmel (1955) that in groups a process of "objectivation" occurs that moves group members beyond a simple concern of self-interests to a situation where individual claims are transformed into "superindividual" concerns. In our terminology, individuals move from the individual to the group pole of our continuum. This may occur under two conditions: (1) when group members feel that they have to act as representatives of their groups; and (2) when group members give their actions a tone of moral righteousness. These conditions are assumed to compel people into greater intransigence and selfless action to further the achievement of group goals (Austin 1979, p. 127).

In this view, the feeling of group belongingness, combined with a sense of moral justification would intensify the social conflict between social groups. If groups attach greater weight to their group interests than individuals would to their own selfish interests, they would experience a greater sense of frustration than individuals about the illegitimate action of the other party which thwarts them in reaching their goals (Brown and Herrnstein 1975). As a consequence, individuals acting as representatives of their group, will react with greater angry aggression to the norm violation of the other party than single individuals would do acting on their own.

One possible implication of this view is that intra-group interaction is not a necessary condition for the greater aggression of groups than of individuals. We were very doubtful about this assumption. According to our enhancement hypothesis (Rabbie and Lodewijkx 1985), derived from our model, intra-group interaction is a necessary and sufficient condition for the greater aggression of groups than of individuals. Consistent with our enhancement hypothesis, the intra-group interaction is supposed to stimulate the psychological orientations and consequent behavior that is already present in individuals prior to the group discussion. In the absence of intra-group interaction no differences between individuals and group representative are to be expected. An experiment was conducted to obtain more information about this issue (Rabbie and Lodewijkx 1987).

In most research in which conflict behavior of individuals and groups are compared with each other, individuals are paired against other individuals, and groups are paired with other groups. This makes it difficult to ascertain whether the greater aggression of groups than of individuals can be attributed to the membership or nonmembership of the actor, or to the membership or nonmembership of the opponent (Rabbie and Horwitz 1982).

In an effort to disentangle these confounding factors, Rabbie and Lodewijkx (1987) designed an experiment in which subjects were randomly either categorized or not categorized, as members of a "Blue" or "Green" group (Rabbie and Horwitz 1969). In the *Group condition* they were urged to act as representatives of their group and to maximize their group's interests as well as they could. In the *Individual condition* they were urged to maximize their own individual self-interests. In both conditions the subjects were seated in separate cubicles so they could not see or hear each other. Thus only coaction but no intra-group interaction was permitted in the group condition. In the individual condition the subjects were acting on their own. From time to time they could see their own and the other (programmed) opponent on a TV screen in front of them. To strengthen their identification with the ingroup, the individual representatives in the group conditions were told that after the end of the experiment they were asked to divide their earnings among themselves. Obviously these instructions were omitted in the individual conditions.

Regardless whether or not they were categorized as group members they faced another individual or a group member who they believed was also instructed to maximize respectively, their individual or group interests. In this way four conditions were created. In the symmetric group-to-group conditions the members of the Blue group faced programmed opponents of Green group members who could be seen on a TV screen in front of them. In the symmetric inter-individual condition noncategorized individuals were paired with a noncategorized individual opponent. In the asymmetric group-individual and individual-group conditions, group members were faced with another individual or an individual faced another group.

Our question was whether more aggression would occur in symmetrical inter-personal or intergroup relations as compared with the asymmetrical relationships. We indicated earlier, discussing the results of the Lodewijkx and Rabbie (1992) experiment, that in symmetrical intergroup relations (interacting) groups, particularly dyads, are likely to follow a more cooperative TFT strategy in a PDG than individuals in symmetric interpersonal relationships. This willingness to cooperate was already reflected in their intentions and expectations indicated prior to the intergroup interactions but also in their actual intergroup behavior. In general it has been shown that similarity or symmetry in a relationship elicits more cooperative orientations in oneself and expectations about the other that he or she is willing to cooperate as well, than in dissimilar or asymmetric relationships (Pruitt and Kimmel 1977; Rabbie 1991a). If a similar, symmetric opponent evokes a stronger expectation of mutual cooperation than a dissimilar or asymmetric adversary, it can be assumed that people would feel more disappointed and frustrated by the deceitful competitive action of the other and hence would react with more aggression to this norm violation in symmetric rather than in asymmetric relationships.

It is difficult to predict whether the norm violation would arouse more or less aggression when people are faced with an individual or a group opponent. It has been proposed that as soon we see people as members or "parts" of a group or social category, the more we tend to "depersonalize" (Horwitz and Rabbie 1982) or even "dehumanize" them (Bandura 1983; Struch and Schwartz 1989). The greater the depersonalization, the less reservations and inhibitions we have to punish the others for their norm violation (Zimbardo 1970; Bandura 1983). On the other hand, we have also speculated that if an individual opponent can be held solely accountable for his or her deceptive behavior there are no extenuating circumstances, as in the case of a group representative, who deceives the other partner on behalf of the group (Rabbie 1982b). Since both hypotheses seemed equally plausible no specific predictions were made.

Consistent with our expectations we found no difference in aggression between individuals and group members. There is thus no support for the hypothesis of Simmel (1955) and others that feelings of group belongingness and a sense of moral righteousness in and of itself would make representatives of groups more aggressive than individuals who are only concerned with maximizing their egoistic self-interests. In many other experiments we have shown that solely when intra-group interaction is allowed, groups react with more aggression to the norm violation of another party than individuals do (e.g., Rabbie and Horwitz 1982; Rabbie and Lodewijkx 1985, 1991; Lodewijkx 1989; Goldenbeld 1992; Rabbie et al. 1992). Apparently, when intra-group interaction is permitted, groups as compared to individuals, become more convinced of the inherent morality of their aggressive actions and are more inclined to use coercive aggression in an attempt to influence the behavior of the other party in the desired direction (Rabbie and Lodewijkx 1987; Janis 1982).

An individual opponent seems to elicit more aggression than a group opponent, regardless whether the actor is a group member or a single individual. This result suggests that, in the absence of mitigating circumstances, a selfish individual is more strongly punished for his or her norm violation than an individual, who deceives one's opponent in the interest of his or her group (Rabbie 1982b).

In agreement with our hypothesis, in symmetric interpersonal and intergroup relationships more aggression was observed than in asymmetric group-individual or individual-group relationship. Apparently, the unexpected deception of a similar opponent, "who is just like me," arouses more aggression than the norm violation of a dissimilar or asymmetric opponent, who is expected to react differently anyway.

As expected, the stronger the cooperative orientations of the subjects during the game; the more intense the aggression to the norm violation of the deceitful and competitive behavior of the opponent. It appears that cooperative subjects who trusted the other to cooperate as well, felt more disappointed and frustrated by the unexpected and devious move of the other party than subjects who were less inclined to cooperate with the other party. In line with deprivation theories (e.g., Gurr 1970), a significant positive relationship was found between the loss the subjects incurred and the aggressive response to the other party.

These findings indicate that even in laboratory situations, the history of the conflict, and the orientations and behavior which are associated with that past, should be taken into account to explain the intensity of current aggression between individuals and groups.

CONCLUSIONS AND IMPLICATIONS

In this paper we have argued that intergroup behavior should be studied at different levels of analysis. Although the psychological cognitive intra-individual approaches to intergroup relations have been very useful and productive (Messick and Mackie 1989), these approaches alone are insufficient to understand and predict the complexities of intergroup conflict and aggression as they occur, for example, in former Yugoslavia, South Africa, and Somalia. In these extreme instances of group aggression, a multidisciplinary approach is required (Rabbie 1989, 1991, 1992).

Our individual-group continuum bears a close resemblance to the individualistic-collectivistic dimension that has been used to characterize different cultures (Hofstede 1980; Triandis 1989; Smith and Bond 1993) and different groups (Hinkle and Brown 1990). Our studies indicate that within a highly individualistic culture like the Netherlands (Hofstede 1980), it is possible to vary the degree of "interdependence and connectedness" (Markus and Kitayama 1991) experimentally.

Although we have found that prior dispositions do make a difference in competitive and aggressive behavior of individuals and groups, we have been more impressed by the impact of the internal and external social environment on the expression of conflict and aggression. In a recent experiment, Goldenbeld (1992) has tried to demonstrate "the power of the situation" (Ross and Nisbett 1991) by introducing a confederate in a dyad who was required to act either as an aggressive or nonaggressive role model (Bandura 1983). For the other member in the dyad he or she was just another fellow-subject just like oneself. It turned out that the role model explained more than 56 percent in the variance of the aggressive behavior of the naive subjects. However, the subjects did not seem to be aware of the massive influence of his or her fellow subject. On our questionnaire they indicated that the other member in the dyad had exerted very little or no influence on their orientations and behavior. It is of course very difficult to assess the degree to which they were really unaware of the influence of the other member. People in our individualistic society like to see themselves as independent "free agents" who are not so much influenced by the actions of others in their group (Markus and Kitayama 1991). Perhaps, for the same reasons, they do not want to admit their dependence on others to the experimenter.

In the same experiment the subjects expected to be held accountable (or not) for their aggressive behavior. In the *accountability condition* they expected to be interviewed about their behavior on the basis of a videorecording that had been made of their interactions within and between the groups involved. In the *no-accountability condition* these instructions were omitted. The accountability manipulation had a significant effect in reducing the aggression of the subjects, but compared with the effects of the role model this effect was very small and explained about 2 percent of the additional variance in aggressive behavior. These results suggest that the actions of a fellow group member in the internal social environment of the dyad has a much greater impact in inhibiting or enhancing aggression than an external authority is likely to have.

In our aggression experiments it was stressed that it was completely up to the subjects whether or not they would use the white noise apparatus to punish the other party for the norm violation. This emphasis on the voluntary character of the aggressive response is quite different from the situation that was created for the subjects in the classic obedience experiments of Milgram (1974). In these studies the subjects were ordered to punish the "learner" for the errors he or she had made in the memory task. In the teacher-learning paradigm and in other laboratory procedures to induce aggression it is more or less taken for granted that the subjects will use the aggression machine (Baron 1977; Krebs and Miller 1985).

Despite the voluntary character of the aggressive response not more than 13 percent of our subjects in our PDG research refused to use the noise apparatus (Lodewijkx 1989). However, we should not overdramatize this finding. Sometimes the intensity of the noise was so low, especially among women, that the noise could barely qualify as physical aggression. On the other hand, on the average, men and woman administered noise levels to the other party that were higher than the noises they described as very unpleasant and almost unbearable.

We were successful in inducing angry aggression in individuals and groups: the higher the intensity of the noise the more subjects rated their own behavior as motivated by anger, irritation, aggression, and indignation. They also indicated more often that the other had deserved his or her punishment (Rabbie and Lodewijkx 1987). There was also a clear relationship between the expression of physical and verbal aggression. The higher the intensity of the noise, the more the group members, especially men, used four letter words and other abusive expressions to characterize the norm violation of the other party (Goldenbeld 1992).

It is interesting to note that many subjects did not consider their behavior as aggressive. They felt that the other got what he deserved. Thus they seemed to consider their behavior as a legitimate punishment of the unacceptable behavior of the other. It can be seen as an attempt to restore the social order that had existed between the two parties before the norm violation of the other party occurred. Possibly, it also reflects a divergence in perspectives between the aggressor and the victim. Mummendy (1984) has shown that people who are requested to identify with an aggressor are less likely to characterize the behavior of a model as aggressive than a person who is asked to take the perspective of the victim. In his or her view, the other is always seen as more aggressive than one self.

One of the most striking and surprising findings of our aggression research has been, despite the many main and interaction effects we have found as a consequence of our experimental manipulations, that groups differ strongly among each other in their aggressive behavior. In fact, the amount of variance in aggressive behavior explained by the "between groups factor" exceeds by far the variation in aggression accounted for by our experimental variables. Apparently, in a few hours, people in laboratory groups construct meaning structures, (sub)cultures or different social climates among each other, which determine the intensity of the conflict and the consequent aggressive behavior. It is one of the main challenges of future research on group conflict and aggression to trace the development of these different collective meaning structures, so that one day we may understand why some groups become more aggressive than other groups.

ACKNOWLEDGMENTS

This chapter is dedicated to the memory of Henry Tajfel who died in 1982 and to Murray Horwitz who passed away in 1991. Both researchers, each one in their own way, have made essential contributions to our thinking about intra- and intergroup relations. The authors would like to thank Lieuwe Visser, Jos van Oostrum, Jan Schot, Jef Syroit, Pawel Mlicki, and many other coworkers and students for their assistance in the research reported in this paper. They are also indebted to the Netherlands Organization for the Advancement of Research (NWO) for their financial support of this research.

REFERENCES

Austin, W. 1979. "Justice, Freedom and Self-interests in Intergroup Conflict." In *The Social Psychology of Intergroup Relations*, edited by W. G. Austin and S. Worchel. Monterey, CA: Brooks/Cole.

Axelrod, R. 1984. *The Evolution of Cooperation*. New York: Basic Book Inc.

Azjen, I. and M. Fishbein. 1980. *Understanding Attitudes and Predicting Social Behavior*. Englewood Cliffs, NJ: Prentice Hall.

Ajzen, I. 1988. *Attitudes, Personality and Behavior*. Milton Keynes: Open University Press.

Bandura, A. 1983. "Psychological Mechanisms of Aggression." In *Aggression, Theoretical and Empirical Reviews*, Volume 1, edited by R. G. Green and E. L. Donnerstein. New York: Academic Press.

Baron, R. A. and R. J. Eggleston. 1972. "Performance on the "Aggression Machine": Motivation to Help or Harm?" *Psychonomic Science* 26:321–322.

Baron, R. A. 1977. *Human Aggression*. New York: Plenum.

Bekkers, F. 1977. "Threatened Leadership and Intergroup Conflicts." *Journal of Peace Research* 3, 14: 223–237.

Branthwaite, A., S. Doyle, and N. Lightbown. 1979. "The Balance Between Fairness and Discrimination." *European Journal of Social Psychology* 9:149–163.

Brown, R. and H. J. Herrnstein. 1975. *Psychology*. London: Methuen.

Brown, R. 1986. *Social Psychology* (2nd Ed). New York: Free Press.

Buss, A. H. 1961. *The Psychology of Aggression*. New York: Wiley.

Cartwright, D. 1968. "The Nature of Group Cohesiveness." In *Group Dynamics*, edited by D. Cartwright and A. Zander. London: Tavistock.

Carver, C. S. and M. Scheier. 1981. *Attention and Self Regulation: A Control Theory Approach to Human Behavior*. New York: Springer.

Colman, A. 1982. *Game Theory and Experimental Games. The Study of Strategic Interaction*. Oxford, Pergamon Press.

Chin, M. G. and C. G. McClintock. 1993. "The Effects of Intergroup Discrimination and Social Values on Level of Self-Esteem in the MGP." *European Journal of Social Psychology* 23:63–75.

Crosby, F. 1976. "A Model of Egoistical Relative Deprivation." *Psychological Review* 83:85–113.

Deutsch, M. 1973. *The Resolution of Conflict: Constructive and Destructive Processes*. New Haven: Yale University Press.

———. 1982. "Interdependence and Psychological Orientation." In *Cooperation and Helping Behavior*, edited by V. J. Derlega and J. Grzelak. New York: Academic Press.

Doise, W. 1986. *Levels of Explanation in Social Psychology*. Cambridge: Cambridge University Press.

———. 1986. "Individual and Social Identities in Intergroup Relations." *European Journal of Social Psychology* 18:99–111.

Dollard, J., L. Dobb, N. Miller, O. H. Mowrer, and R. R. Sears. 1939. *Frustration and Aggression*. New Haven, CT: Yale University Press.

Eagly, A. H. 1987. *Sex Differences in Social Behavior: A Social-role Interpretation*. Hillsdale, NJ: Erlbaum.

Ferguson, T. J. and B. G. Rule. 1983. "An Attributional Perspective on Anger and Aggression." In *Aggression: Theoretical and Empirical Reviews*, Vol. 1, edited by R. G. Geen and E. I. Donnerstein. New York: Academy Press.

Feshbach, S. 1964. "The Function of Aggression and the Regulation of Aggressive Drive." *Psychological Review* 71:257–272.

Gerard, H. B. and J. M. Rabbie. 1961. "Fear and Social Comparison." *Journal of Abnormal and Social Psychology* 62:586–592.

Goldenbeld, C. 1992. "*Aggression After Provocation*." Dissertation, University of Utrecht.

Gouldner, A. W. 1960. "The Norm of Reciprocity: A Preliminary Statement." *American Sociological Review* 25:161–178.

Gurr, T. 1970. *Why Men Rebel*. Princeton, NJ: Princeton University Press.

Hinkle, S. and R. J. Brown. 1990. "Intergroup Comparisons and Social Identity: Some Links and Lacunae." Pp. 48–70 in *Social Identity Theory: Constructive and Critical Advances*, edited by D. Abrams and M. Hogg. Hemel Hemstead: Harvester Whatsheaf/New York: Springer Verlag.

Hofstede, G. 1980. *Culture's Consequences: International Differences in Work-related Values*. Beverley Hills, CA: Sage.

Hogg, M. A. and D. Abrams. 1988. *Social Identification: A Social Psychology of Intergroup Relations and Group Processes*. London and New York: Routledge.

Horwitz, M. and J. M. Rabbie. 1982. "Individuality and Membership in the Intergroup System." In *Social Identity and Intergroup Relations*, edited by H. Tajfel. Cambridge University Press: Editions de la Maison des Sciences de l'Homme.

———. 1989. "Stereotypes of Groups, Group Members and Individuals in Categories: A Differential Analysis of Different Phenomena." In *Stereotyping and Prejudice: Changing Conceptions*, edited by D. Bar-Tal et al. New York: Springer Verlag.

Jaffe, Y. and Y. Yinon. 1983. "Collective Aggression: The Group-individual Paradigm in the Study of Collective Social Behavior." In *Small Groups and Social Interaction*, Volume 1, edited by H. H. Blumberg, A. P. Hare, V. Kent, and M. Davis. New York: Wiley.

Janis, I. L. 1982. *Groupthink*. Boston: Houghton Mifflin.

Kelley, H. H. and A. Stahelski. 1970. "The Social Interaction Basis of Cooperators' and Competitors' Beliefs About Others." *Journal of Personality and Social Psychology* 16:66–91.

Kelley, H. H. and J. W. Thibaut. 1978. *Interpersonal Relations: A Theory of Interdependence*. New York: Wiley, Inter Science.

Krebs, D. L. and D. T. Miller. 1985. "Altruism and Aggression." Pp. 1–77 in *The Handbook of Social Psychology*, Volume II, edited by G. Lindzey and E. Aronson. New York: Random House.

Kroon, B. R. 1992. "Effects of Accountability on Groupthink and Intergroup Relations: Laboratory and Field Studies." Dissertation, University of Utrecht.

Kroon, B. R., D. van Kreveld, and J. M. Rabbie. 1991. "Police Intervention in Riots: The Role of Accountability and Group Norms." *Journal of Community and Applied Social Psychology* 1:249–267.

———. 1992a. "Police Intervention in Riots: A Reply to Chief Superintendent Tom Wiliamson, Commander Tony Row, and Dr. Steve Reicher." *Journal of Community and Applied Social Psychology* 2:73–75.

———. 1992b. "Group Versus Individual Decision Making." *Small Group Research* 23, 4:427–458.

LeBon, G. 1895. *Psychologie des Foules. (The Crowd)*: London: T. Fisher, Unwin.

LeVine, R. A. and D. T. Campbell. 1972. *Ethnocentrism: Theories of Conflict, Ethnic Attitudes and Group Behavior*. New York: Wiley.

Lewin, K. 1948. *Resolving Social Conflicts*. New York: Harper and Row.

———. 1952. *Field Theory in Social Science*, edited by D. Cartwright. New York: Harper & Bros.

Lodewijkx, H. 1989. *"Conflict and Aggression Between Groups and Individuals: Toward a Behavioral Interaction Model."* Dissertation, University of Utrecht.

———. 1992. "Evocation of Matrix Transformations in the MGP by Accountability and Groupboundaries." Paper presented at the seminar on intergroup relations, Amsterdam.

Lodewijkx, H. F. M., P. P. Mlicki, J. C. Schot, and J. M. Rabbie. 1991. *"Interdependence and the Minimal Group Paradigm."* Paper presented at the Social Value Conference.

Lodewijkx, H. F. M. and J. M. Rabbie. 1992. "Group-centered and Self-centered Behavior in Intergroup Relations." *International Journal of Psychology* 27, 3–4:267.

Lodewijkx, H. F. M. and N. Akkersdijk. 1993. *Initiation in a Dutch Sorority*. Unpublished manuscript: Utrecht University.

Markus, H. and S. Kitayama. 1991. "Culture and the Self: Implications for Cognition, Emotion and Motivation." *Psychological Review* 98:224–253.

Merton, R. K. 1968. *Social Theory and Social Structure*. Glencoe, IL: Free Press.

McClintock, C. G. 1972. "Game Behavior and Social Motivation in Interpersonal Settings." Pp. 271–297 in *Experimental Social Psychology*, edited by C.G. McClintock. New York: Rinehart and Winston.

Messick, D. M. and D. M. Mackie. 1989. "Intergroup Relations." *Annual Review of Psychology* 40:45–81.

Milgram, S. 1974. *Obedience to Authority: An Experimental View*. New York: Harper and Row.

Miller, G. A., E. Galanter, and K. H. Pribram. 1960. *Plans and the Structure of Behavior.* New York: Holt, Rinehart & Winston.

Mlicki, P. P. 1993. *"Us and Them": Effects of Categorization and Interdependence on Differentiation Between Categories and Groups."* Dissertation, University of Utrecht.

Mummendy, A. 1984. *Social Psychology of Aggression. From Individual Behavior to Social Interaction.* Heidelberg: Springer Verlag.

Ng, C. 1984. "Equity, and Social Categorization Effects of Intergroup Allocations or Rewards." *British Journal of Social Psychology* 23:165–172.

———. 1985. "Equity, Intergroup Bias, and Interpersonal Bias in Reward Allocation." *European Journal of Social Psychology* 16:239–255.

Petty, R. E. and J. T. Cacioppo. 1986. *Communication and Persuasion: Central and Peripheral Routes to Attitude Change*. New York: Springer Verlag.

Platow, M. J., C. G. McClintock, and W. B. G. Liebrand. 1990. "Predicting Intergroup Fairness and Ingroup Bias in the MGP." *European Journal of Social Psychology* 20:221–239.

Porter, L. W. and E. Lawler. 1968. *Managerial Attitudes and Performance*. Homewood, IL: Irwin.

Pruitt, D. G. and M. J. Kimmel. 1977. "Twenty Years of Experimental Gaming: Critique, Synthesis, and Suggestions for the Future." *Annual Review of Psychology* 28:363–392.

Pruitt, D. G. and J. C. Rubin. 1986. *Social Conflict: Escalation, Stalemate and Settlement*. New York: Random House.

Rabbie, J. M. 1963. "Differential Preference for Companionship Under Threat." *Journal of Abnormal and Social Psychology* 67:643–648.

———. 1964. *"Ingroup-outgroup Differentiation Under Minimal Social Conditions."* Second European Conference of Experimental Social Psychology, Frascati, Italy.

———. 1974. "Effects of Expected Intergroup Competition and Cooperation." Paper presented to the Symposium on the Development and Maintenance of Intergroup Bias. Annual Convention of the A.P.A., New Orleans.

———. 1982a. "The Effects of Intergroup Competition on Intra-group and Intergroup Relationships." In *Cooperation and Helping Behavior: Theories and Research*, edited by V. J. Derlega and J. Grzelak. New York: Academic Press.

———. 1982b. *"Are Groups More Aggressive than Individuals?"* Presented at the Annual Conference of the Social Psychology Section of the British Psychological Society.

———. 1987. "Armed Conflicts: Toward a Behavioral Interaction Model." In *European Psychologists for Peace. Proceedings of the Congress in Helsinki, 1986*, edited by J. von Wright, K. Helkema, and A.M. Pirtilla-Backman.

———. 1989. "Group Processes and Stimulants of Aggression." In *Aggression and War: Their Biological and Social Bases*, edited by J. Groebel and R.H. Hinde. Cambridge: Cambridge University Press.

———. 1990. "Aggressive Conflicts Between Individuals and Groups: A Social Psychological Approach." Pp. 175–236 in *War: Multidisciplinary Contributions*, edited by J.A.R. A. M. van Hooff, G. Benthem van den Bergh, and J. M. Rabbie. Hoogezand: Stuberg.

———. 1991a. "Determinants of Instrumental Intra-Group Cooperation." In *Cooperation and Prosocial Behavior*, edited by J. Groebel and R. Hinde. Cambridge: University of Cambridge Press.

———. 1991b. "A Behavioral Interaction Model: A Theoretical Framework for Studying Terrorism." *Terrorism and Political Violence* 3,4:133–162.

————. 1992. "Effects of Intra-group Cooperation and Intergroup Competition on Ingroup-outgroup Differentiation." In *Cooperation in Conflict: Coalitions and Alliances in Animals and Humans,* edited by A. Harcourt and F. de Waal. Oxford: Oxford University Press.

————. 1993a. "A Behavioral Interaction Model: Towards an Integrative Framework for Studying Intra- and Intergroup Relations." Pp. 86–108 in *Conflict and Social Psychology,* edited by K. Larson. Beverley Hills, CA: PRIO/Sage.

————. 1993b. "Determinants of Ingroup Cohesion and Outgroup Hostility." *International Journal of Group Tensions* 23(4):309–328.

Rabbie, J. M. and F. Bekkers. 1976. "Threatened Leadership and Intergroup Competition." *European Journal of Social Psychology* 31:269–283.

Rabbie, J. M., F. Benoist, H. Oosterbaan, and L. Visser. 1974. "Differential Power and Effects of Expected Competitive and Cooperative Intergroup Interaction on Intra-group and Outgroup Attitudes." *Journal of Personality and Social Psychology* 30:46–56.

Rabbie, J. M. and J. H. C. de Brey. 1971. "The Anticipation and Competition Under Private and Public Conditions." *International Journal of Group Tensions* 1:230–252.

Rabbie, J. M., C. Goldenbeld, and H. F. M. Lodewijkx. 1992. "Sex Differences in Conflict and Aggression in Individual and Group Settings. Pp. 217–228 in *Of Mice and Women: Aspects of Female Aggression,* edited by K. Björquist and P. Niemellä. New York: Academic Press.

Rabbie, J. M. and M. Horwitz. 1969. The Arousal of Ingroup-Outgroup Bias by a Chance Win or Loss. *Journal of Personality and Social Psychology* 69:223–228.

————. 1982. "Conflict and Aggression Among Individuals and Groups." In *Proceedings of the XXIInd International Congress of Psychology, Leipzig, DDR, No. 8,* edited by H. Hiebsch, H. Brandstätter, and H. Kelley.

————. 1988. Categories Versus Groups as Explanatory Concepts in Intergroup Relations. *European Journal of Social Psychology* 18:117–123.

Rabbie, J. M. and K. Huygen. 1974. "Internal Disagreements and their Effects on Attitudes Towards In- and Outgroup." *International Journal of Group Tensions* 4:222–246.

Rabbie, J. M. and H. F. M. Lodewijkx. 1985. "The Enhancement of Competition and Aggression in Individuals and Groups." In *Social/Ecological Psychology and the Psychology of Women,* edited by F. L. Denmark. Amsterdam: North-Holland Publishing Company.

————. 1987. "Individual and Group Aggression." *Current Research on Peace and Violence* 2–3:91–101.

————. 1991. "Aggressive Reactions to Social Injustice by Individuals and Groups: Toward a Behavioral Interaction Model." In *Social Justice in Human Relations,* Volume 1, edited by R. Vermunt and H. Steensma. New York: Plenum Press.

————. 1992. *Instrumental and Relational Cooperation and Competition Along an Individual-group Continuum.* Unpublished manuscript, University of Utrecht.

Rabbie, J. M., H. Lodewijkx, and M. Broeze. 1985. "Individual and Group Aggression Under the Cover of Darkness. P. 140 in *Multidisciplinary Approaches to Conflict and Appeasement in Animals and Man,* edited by L. Moli. Instituto di Zoologica, Universita degi Studi di Parma.

Rabbie, J. M., L. Visser, and van Oostrum. 1982. "Conflict Behavior of Individuals, Dyads and Triads in Mixed-Motive Games." In *Group Decision-making,* edited by H. Brandstätter, J. H. Davis, and G. Stocker-Kreichgauer. London: Academic Press.

Rabbie, J. M., L. Visser, and L. Vernooy. 1976. "Uncertainty of the Environment, Differentiation and Influence Distribution in University Departments." *Nederlands Tijdschrift voor de Psychologie* 31:285–303.

Rabbie, J. M. and L. Visser. 1972. "Bargaining Strength and Group Polarization in Intergroup Negotiations." *European Journal of Social Psychology* 4:401–416.

Rabbie, J. M. and J. C. Schot. 1989. "*Instrumental and Relational Behavior in the Minimal Group Paradigm.*" Paper presented at the first European Congress of Psychology in Amsterdam, The Netherlands, July 7.

———. 1990. "Group Behavior in the Minimal Group Paradigm: Fact or Fiction?" Pp. 251–263 in *European Perspectives in Psychology*, Vol. 3, edited by P. J. D. Drenth, J. A. Sergeant, and R. J. Takens. Chichester: Wiley and Sons.

Rabbie, J. M., J. C. Schot, and L. Visser. 1989. "Social Identity Theory: A Conceptual and Empirical Critique From the Perspective of a Behavioral Interaction Model." *European Journal of Social Psychology* 19:171–202.

Rabbie and Wilkens, G. 1971. Intergroup Competition and Its Effects on Intra-Group and Intergroup Relations. *European Journal of Social Psychology* 1:215–234.

Rapaport, A. and A. M. Chammah. 1965. *Prisoner's Dilemma: A Study in Conflict and Cooperation.* Ann Arbor: University of Michigan Press.

Ross, L. and R. E. Nissbet. 1991. *The Person and the Situation: Perspectives of Social Psychology.* New York: McGraw Hill.

Schachter, S. 1959. *The Psychology of Affiliation.* Stanford, CA: Stanford University Press.

Schopler, J. and C.A. Insko. 1992. "The Discontinuity Effect in Interpersonal and Intergroup Relations: Generality and Mediation." Pp. 121–151 in *European Review of Social Psychology*, Vol. 3, edited by W. Stroebe and M. Hewstone.

Schot, J. C. 1992. *Allocations in the Minimal Group Paradigm.* Dissertation, Utrecht University.

Sherif, M. 1966. *Group Conflict and Co-operation: Their Social Psychology.* London: Routledge & Kegan Paul.

———. 1967. *In Common Predicament.* Boston: Houghton Mifflin Company.

Sherif, M. and C. W. Sherif. 1979. "Research on Intergroup Relations." In *The Social Psychology of Intergroup Relations*, edited by W. C. Austin and S. Worchel. Monterey, CA: Brooks Cole.

———. 1964. *Reference Groups: Exploration into the Conformity and Deviation of Adolescents.* New York: Harper and Row.

Simmel, G. 1955. *Conflict.* New York: Free Press of Glencoe.

Smith P. B. and M. H. Bond. 1993. *Social Psychology Across Cultures: Analyses and Perspectives.* Hemel Hemstead: Harvester Whatsheaf.

Stroebe, W. and C. A. Insko. 1989. "Stereotype, Prejudices and Discrimination: Changing Conceptions in Theory and Research." In *Stereotyping and Prejudice: Changing Conceptions*, edited by D. Bar-Tal et al. New York: Springer Verlag.

Struch, N. and S. H. Schwartz. 1989. "Intergroup Aggression: Its Predictors and Distinctiveness From In-Group Bias." *Journal of Personality and Social Psychology* 56, 3:364–373.

Sumner, W. G. 1906. *Folkways.* New York: Ginn & Co.

Syroit, J. E. E. M. 1991. "Interpersonal and Intergroup Justice: Some Theoretical Considerations." In *Social Justice in Human Relations*, Volume 1, edited by R. Vermunt and H. Steensma. New York: Plenum Press.

Tajfel, H. 1982. "Instrumentality, Identity and Social Comparison." Pp. 483–507 in *Social Identity and Intergroup Relations*, edited by H. Tajfel. Cambridge University Press: Editions de la Maison des Sciences de l'Homme.

Tajfel, H., M. G. Billig, H. P. Bundy, and C. I. Flament. 1971. "Social Categorization and Intergroup Behavior." *European Journal of Social Psychology* 1:149–178.

Tajfel, H. and J. C. Turner. 1979. "An Integrative Theory of Intergroup Conflict." Pp. 33–47 in *The Social Psychology of Intergroup Relations*, edited by W.G. Austin and S. Worchel. Monterey, CA: Brooks Cole.

———. 1986. "The Social Identity Theory of Intergroup Behavior." In *The Social Psychology of Intergroup Relations*, edited by S. Worchel and W. G. Austin. Chicago: Nelson-Hall.

Triandis, H.C. 1989. "The Self and Social Behavior in Different Cultural Contexts." *Psychological Review* 96:506–520.

Turner, J. C. 1980. "Fairness or Discrimination in Intergroup Behavior? A Reply to Branthwaite, Doyle & Lightbown." *European Journal of Social Psychology* 10:131–147.

———. 1981. "The Experimental Social Psychology of Intergroup Behavior." In *Intergroup Behavior*, edited by J. C. Turner and H. Giles. Oxford: Blackwell.

———. 1982. "Towards a Cognitive Redefinition of the Social Group." In *Social Identity and Intergroup Relations*, edited by H. Tajfel. Cambridge: Cambridge University Press.

Turner, J. C., M. A. Hogg, P. J. Oakes, S. D. Reicher, and M. S. Wetherell. 1987. *Rediscovering the Social Group: Self Categorization Theory.* Oxford: Basil Blackwell.

Van Avermaet, E. and C. G. McClintock. 1988. "Intergroup Fairness and Bias in Children." *European Journal of Social Psychology* 18:407–427.

Van Leent, J. A. A. 1964. *Sociology, Psychology and Social Psychology, Their Design, Development and Relationship from a Macro-micro Perspective.* Arnheim: de Haan/van Loghum Slaterus.

Visser, L. 1993. *"Cooperation and Competition among Groups and Individuals."* Unpublished dissertation, Utrecht University.

Weber, M. 1921. *Wirtschaft und Gesellschaft.* Tübingen: J.C.B. Mohr.

With, J. S. and J. M. Rabbie. 1985. "Racist Attitudes about Intimate Relationships between Blacks and Whites in the Netherlands." *Behavior: Journal for Psychology* 13:10–13.

Wilson, W. 1971. "Reciprocation and Other Techniques for Inducing Cooperation in the Prisoner's Dilemma Game." *The Journal of Conflict Resolution* 15:167–195.

Zimbardo, P. G. 1970. "The Human Choice: Individuation, Reason, and Order Versus Deindividuation, Impulse and Chaos." In *Nebraska Symposium on Motivation*, edited by W. J. Arnold and D. Levine. Lincoln: University of Nebraska Press.

COHERENT STRUCTURE THEORY AND DECISION MAKING IN SOCIAL NETWORKS:

A CASE STUDY IN SYSTEMS ISOMORPHY

D. H. Judson and L. N. Gray

ABSTRACT

This paper is a case study in systems isomorphy or systems analogy. We use the theory of coherent structures, originally developed for engineering applications, to analyze decision making in social networks under a variety of structural and normative constraints. We assert two things: (1) That the theory is a useful generalization of much of the work in social choice and coalition theory; and (2) That the theory is broader and more in accord with social network approaches than much social choice theory up to now. To support these theses, we first explore and develop psychological, sociological, and historical interpretations of many of the terms in coherent structure theory as it is applied to decision-making groups. Second, we show that coherent structure theory has implications that are in accord with summaries of empirical studies of group decision making. Third, we show that coherent decision-making

Advances in Group Processes, Volume 11, pages 175–211.
Copyright © 1994 by JAI Press Inc.
All rights of reproduction in any form reserved.
ISBN: 1-55938-857-9

structures encompass a larger (and more important) class of group structures than weighted majority rules. Fourth, we explore the implications of the existence of this larger class of decision rules for social decision theory.

INTRODUCTION

In a recent book on theoretical sociology, Thomas Fararo (1989, pp. 253–254) argued that sociologists should start with a dynamic action model of the individual and then link models of actors to derive social systemic outcomes. Other writers (Coleman 1990, p. 18; Burt 1982, pp. 3–4, 329; Tallman and Gray 1990) have made similar claims. Their claims are particularly important because they see in sociology a need for two theories: (1) a theory of the individual, and (2) a theory linking the individual with their network of social relations.

One distinct strength of individual choice modeling is theoretical precision. Because choice theories are typically cast in mathematical form, their predictions are more easily testable, and hence falsifiable. The result has been a good deal of formal development. However, one major criticism of classical individual choice modeling, which we address in this paper, is its thoroughgoing individualism. Only rarely did classical writers consider what is now known as the problem of social choice. In early work, a decision situation was seen as a collection of strategies from which the individual decision maker selects, as if from a dessert tray. What was ignored was precisely the factors that make decision making sociologically interesting—*social contexts*. In social contexts, rewards and costs experienced by decision makers are not constant, as presumed in individual situations. Instead, they are created by other decision makers within the same social context. That is, the environment in which *one* decision maker functions is that created by the collection of decision makers, and not one that can be easily specified (as by an experimenter) a priori.

CLASSICAL DECISION THEORY AND THE PROBLEM OF SOCIAL CHOICE

Classical decision theory, starting with the seminal work of Bernoulli's (1713) *Ars Conjectandi* (*The Art of Conjecture*), represents each individual as facing an uncertain environment. On this, decision analysts agree. How to represent the uncertain environment, however, is more controversial (Raiffa 1968, pp. xix–xx). Classical decision theory represents "states of nature" as events having fixed probabilities of occurrence unknown to the decision maker, in which the goal of the decision maker is to ascertain with minimal error what these probabilities are. Modern decision theory, in contrast, models an individual's willingness to assign prior subjective probabilities to a fixed set of "states of nature," to incorporate new data resulting from these probabilities, and to generate revised posterior probabilities based on these data. As early as 1847, De Morgan argued that probability should be considered a subjective concept, a "degree of belief." This was echoed in the twentieth century by Keynes (1921), Russell (1948), and Savage (1954). In the course of development of what has come to be called decision analysis, several powerful tools for maximizing expected returns have been developed, notably decision trees, linear programming techniques, and tools for eliciting the necessary subjective probabilities and magnitudes. These tools have influenced such diverse areas as mathematical statistics (e.g., Zellner 1971), the theory of collective choice (Arrow 1951/1963), international bargaining (Wang, Hipel, and Fraser 1988), as well as others (see Watson and Buede 1987, for a review).

Such individual decision theories have not gone unchallenged, however. The development of Game Theory (Von Neumann and Morgenstern 1944) represented a first step toward representing decision making as a social event. In the traditional game theory context, decision makers are represented as choosing among a fixed menu of strategies presuming each decision maker (player) knows that the other(s) will optimize their expected outcomes. Given this model presumption, an optimal solution to many kinds of strategic situations can be derived (see e.g., Shubik 1982). Indeed, game theorists have demonstrated that classical statistical decision problems can be considered as games played against a strategic opponent called "Nature." The problem is then transformed into this: How does one model the strategies of "Nature?" That is, what is the probability that Nature will choose any particular outcome?

What is missing in all these analyses is a representation of decision making *itself* as a social event. That is, modern challenges to the classical decision-making scheme have emphasized that more often groups of decision makers *collectively* attempt to generate solutions. One approach (Raiffa 1968, pp. 233–234) to solving this problem is to let each member contribute to a group "consensus" about probabilities and benefits, and then treating the group consensus as an aggregate individual. Savage went so far as to say that the theory "originally intended to apply to people may also apply to (or may even apply better to) such units as families, corporations, or nations" (1954, p. 8). While such an approach preserves the traditional tools of decision analysis, it does little to represent the process of group consensus. A more "social" solution is to model each individual decision maker in a decision-making team as making individual judgments, in the manner of French (1956). Then, some collective decision-making rule must be specified by which the individual judgments are pooled into a collective judgment. It is precisely the *collective decision-making rule* which shifts attention away from individualistic analyses to an explication of social structure. The goal of this paper is to construct a general theoretical framework for analyzing collective decision making.

SCOPE OF THIS PAPER

To define the generalizability of the theory we are developing, and limit this first attempt to a manageable size, we present the following scope conditions. These conditions broadly define the empirical situations to which the theory applies, and from which observations can be made that would falsify the theory. Further, they are stated in (perhaps excessive) generality to make the theory as broadly applicable as is reasonable. We use the term "researcher" to represent the individual observing the decision-making group.

1. The theory applies only to finite groups of N individuals with the constraint that these N individuals must generate a decision representing their collective judgment.
2. Individuals within the group may or may not be aware of each others' presence; likewise, they may or may not be aware of the structure of the decision-making group. Awareness (or lack thereof) makes no difference to individual decision processes.
3. Decision options, in this theory, are dichotomous, but not symmetric; each individual has the ability to vote *for* an option or to vote *against* it.[1] Though this appears to be a stringent limitation, empirical choices involving a finite set of discrete options can be theoretically represented as a series of dichotomous choices. Legislative amendment voting is the archetypal case.
4. Each member of the group may be assigned a priori a number from zero to one (inclusive) representing their probability of choosing any particular option; that is, each member has a competence value that is either known, assumed, or modeled by the researcher.
5. The decision rule employed by the decision-making group may be consciously chosen, overt and formal, or it may be nonconscious and informal; regardless of the group's knowledge of the decision rule, it is known to the researcher.
6. This theory does not specify *how* coalitions come to exist; rather, it is presumed that the structure is preexisting and known to the researcher. However, it is possible for a researcher to empirically examine an interaction system and determine which coherent structure fits group outcomes best.
7. The purpose of this theory is to predict the probability of a dichotomous outcome given a certain coherent structure and certain individual competencies, not to predict which kind of coherent structure will form.

CONCEPTS IN THE THEORY OF GROUP DECISION MAKING

There are several theoretical strands to the modern emphasis on social decision making. In this section, we discuss four; each of which attempts to explain, theoretically or empirically, the effects of social relationships in group decision making. The four include: (1) The formal theoretical tradition (e.g., game theory); (2) the empirical tradition (e.g., social-psychological research); (3) Individual decision making in social contexts; and (4) coalition research.

The Formal Theoretical Tradition

In the formal theoretical tradition, Grofman (1978; Grofman, Owen, and Feld 1983; Grofman and Feld 1986), Shapley and Grofman (1984), Nitzan and Paroush (1984a,b,c), and others, have focused on deriving formal decision rules that translate fixed a priori individual judgmental competencies into optimal group judgments. The problem of maximizing group decisional competence was given its original form in de Condorcet (1785), in which N individuals with equal decisional competence[2] attempt to maximize their collective probability of making a correct choice. (Throughout this paper, "competence" will be defined as the probability of an individual or group making an arbitrary "correct" choice.) The resulting theorem, known as the Condorcet Jury Theorem (see Grofman and Owen 1986; Batchelder and Romney 1986) states the following:

Let p be the probability that any individual decision maker among N decision makers will make the correct choice in a dichotomous decision situation, and let m be a majority ($m = (N+1)/2$) for odd N. Let P_N be the probability that a majority of the group will pick the correct alternative. If voter choices are mutually independent, then:

$$P_N = \sum_{h=m}^{N} \binom{N}{h} p^h (1 - p)^{N-h}$$

If $1 > p > 1/2$, then P_N is monotonically increasing in N and $\lim_{N \to \infty} P_N = 1$;
If $0 < p < 1/2$, then P_N is monotonically decreasing in N and $\lim_{N \to \infty} P_N = 0$;
If $p = 1/2$, then $P_N = 1/2$ for all N.

The Condorcet model was developed in and reflective of the democratic fervor that was revolutionary France. It demonstrated that, if each and every voter were more competent than a coin flip (i.e., they had better than 50–50 odds of making a correct choice), then the democratic majority would inexorably lead toward perfect competence (see Young 1988; Grofman and Feld 1988, for further analyses of Condorcet's theorem).

Theoreticians have moved away from the original Condorcet model of equally competent decision makers, however, toward a more general conception ofthe problem. The most important recent result, due to Shapley and Grofman (1984), and Nitzan and Paroush (1982), is the following result, stated infor- mally:

> Suppose that we have a group of $I = 1,2, \ldots, n$ decision makers, each voting for one of two choices ($x_i = -1$ or $x_i = 1$), with a fixed and known probability p_i of making the "correct" choice. Then the decision rule which maximizes the likelihood of the group making the "correct" choice is a weighted majority rule of the form[3]: Sgn ($w_1x_1+w_2x_2+\ldots +w_nx_n$), where $w_i = \ln(p_i/1-p_i)$.

This result indicates that to maximize the use of each individual's competency, the group should use a weighted majority rule as their decision rule. Essentially, the group should add up the weights (w_i) times the votes (x_i), and if the sum is greater than zero, choose one as the group's decision. If the sum is less than zero, choose −1 as the group's decision. Interestingly, the resulting optimal weights for the weighted majority rule correspond to logits, $\ln(p_i /1-p_i)$, and individuals with competencies less than .5 should receive negative weights (that is, if my compe- tency is low, and I vote for strategy A, the group should be more inclined toward option B!). Extensions and elaborations of this result are available in Nitzan and Paroush (1984a, 1984b, 1984c), and Karotkin, Nitzan, and Paroush (1988). The analysis of group judgmental competence is interesting on theoretical grounds alone, but has been criticized for its apparent lack of connection with empirical data (Hastie 1986).

The Empirical Tradition

A related strand of research, focusing on experimental elaboration and empirical applications, is the group decision rules studies (see e.g., Miller 1989). Such studies have focused on issues such as failure to reach a decision given a particular decision rule, probability and extent of compromise, and empirical quality of decisions. In contrast to developing models for optimizing group decision making or modeling the process of group decision making, this research tradition instead has explored the social psychological consequences of choosing a particular formal decision rule in empirical contexts. The central group decision rules considered are unanimity rule (all must agree), majority rule (a specified majority must agree), and various forms of veto rules (where one individual has the unilateral ability to block a decision but not to pass a decision.)

The "most frequently reported" (Miller 1989, p. 330) finding in this research is that *unanimity rule is more likely to result in failure to reach a decision than most types of majority rule*. This has been found both in jury and non-jury studies (e.g., Foss 1981; Hastie, Penrod, and Pennington 1984; Kerr, Atkin, Stasser, Meek, Holt, and Davis 1976). However, when a decision is not dichotomous (e.g., guilty/not guilty), it has been found that unanimity rules lead to greater levels of compromise than various majority rules (Harnett 1967; Miller 1985).

In terms of decisional quality, several studies have determined that unanimity rule, despite difficulties in generating conclusions, generally results in higher quality conclusions (Holloman and Hendrick 1972; Bower 1965a, 1965b). Bower argued that unanimity rule was superior due to the requirement that individuals take account of each other's unique information. Thompson, Mannix, and Bazerman (1988) suggest a similar mechanism: Unanimity rule systems must take into account not only the direction of preference but the strength of preference, and this leads to a more "integrative" or "maximizing" solution. Thus we summarize the following important generalization from empirical studies: It appears that unanimity rule is a better strategy, in certain situations, than are majority rules.

In his 1989 review of the group decision rule literature, Miller argued that two characteristics of group decision rules were particularly important. The first, called *strictness*, refers to the "extent to which group members must express similar preferences in order for the group to reach a decision" (Miller 1989, p. 329). Thus, for example, an unanimity rule is stricter than a dictator rule, because the former requires all votes be equivalent, while the latter requires only that one person's opinion be taken into account. The second characteristic, called *distribution of power*, represents the distribution of individual influences on the group decision. A decision rule such as dictator rule concentrates power at one point, while unanimity distributes it equally among all members. While these two characteristics are related, they tap different dimensions of decision rules. For example, both equal-vote majority and unanimity rules distribute voting power evenly among all members, but the latter is more strict than the former.

Individual Decision Making in Social Contexts

Another related strand of research is that of individual decision making as it is applied to social contexts (Gray and Stafford 1988). This research program has applied an individual decision-making model known as the Satisfaction-Balance Model (Gray and Tallman 1984) to situations similar to group decision rules studies. The Satisfaction-Balance Model is based on the matching law of behavioral psychology (Herrnstein 1970, 1974), and notions of probability matching (Gray, Stafford, and Tallman 1991). This model argues that individuals match their odds of choosing an option with the weighted ratios of expected reward and expected punishment (see Gray and Tallman 1984).

Applying this model to individuals who then negotiate in a group decision-making context has resulted in the derivation of probabilities of *groups* making certain decisions under conditions where each *individual* chooses according to the model. The primary theoretical and empirical result of Gray and Stafford (1988) is the following: Suppose that a group of individuals is placed in a dichotomous decision situation. Suppose further that each outcome has certain fixed probabilities of reward/cost and fixed magnitudes of reward/cost. Finally, suppose that each individual evaluates the probabilities and magnitudes of reward/cost according to the Satisfaction Balance Model. *Then a unanimity group decision rule results in a greater probability of the group choosing the more favorable option than any individual probability alone.*

This result parallels the empirical results discovered under the rubric of "risky shift" (Moscovici and Zavalloni 1969; Vinokur 1971; Dion, Baron, and Miller 1978; Moscovici 1985) and is also parallel to the result (discussed here) that unanimity rule is a better strategy than various forms of majority rule. From the perspective of the matching law of individual psychology, this result suggests that groups will be *less* likely to match and *more* likely to maximize their rewards than individuals. In their empirical examination of this model, Gray and Stafford found substantial support for the model when they compared individuals and two-person groups.

Coalition Research

A fourth research tradition of direct relevance for the study of group decision making is that of coalition formation. Gamson (1964, p. 82; see also Komorita 1974; Komorita and Kravitz 1983) defined the term coalition as "the joint use of resources to determine the outcome of a decision," while a resource was defined as "some weight controlled by the participants such that some critical quantity of these weights is necessary and sufficient to determine the decision." The study of coalition formation in groups has proceeded rapidly, due in part to a relative abundance of good historical data on coalitions in parliamentary bodies (e.g., De Swaan 1973, 1985), and due in part to its suitability for laboratory study. Indeed, laboratory coalition research is more directly isomorphic to real world coalition formation than many kinds of social psychological studies. Furthermore, the existence of metric weights (votes or resources) in such situations has encouraged a wide variety of specific formal models of coalition formation. The analysis by Georg Simmel (1950) of social relations in the dyad and triad served as a catalyst for the pioneering studies of Mills (1953), Caplow (1956), and Gamson (1961a, 1961b).

Coalition formation research has progressed to a point where experimental or historical situations can be represented in a standard form. For example, suppose four persons play a coalition game where they must pool their resources to determine an outcome. If nine "votes" are required to win, and persons A,B,C, and D have eight, three, three, and three votes respectively, the standard form representation of this game is 9(8–3–3–3). This is important because such a representation casts the game as one using a *weighted majority rule*, where majority is defined as 9 votes, and weights 8, 3, 3, and 3 are assigned to parties A, B, C, and D.

SYSTEMS ISOMORPHY:
THE THEORY OF COHERENT STRUCTURES

The study of group decision making and social choice has developed with little regard for parallel developments in social structural analysis. Yet, it is immediately apparent from our discussion that the analysis of group decision making cries out for a model of social structure (Wellman 1988). Where can a model of social structure be found?

In 1961, Birnbaum, Esary, and Saunders published "Multicomponent Systems and Structures, and Their Reliability." Their purpose in that paper was to develop a formal structure in which the reliability of systems of physical (i.e., mechanical, electronic) components could be analyzed. However, as we will see, their theory of coherent structures requires only well-defined components and relations between the components, not necessarily well-defined *physical* components and *physical* relations. We will use this theory, with modifications appropriate to decision-making social structures, as the foundation for the models developed here.

Structural models have much intuitive appeal; they formally represent the world in a way that matches many researchers' intuitive conceptions of the operation of social structure (Berkowitz 1988). They provide a foundation for linking research on individual decision making with research on group decision making. Further, structural models are models of *larger* networks of individuals and groups; instead of limiting decision making to individuals, or even small groups, structural models extend that conception to large networks of persons, even to organizational decision making. Finally, several researchers have bemoaned the lack of theoretical development (as opposed to methodological development) in the social network approach (e.g., Granovetter 1979; Freeman 1989); a link between structural models and a well established model of the individual will have the beneficial effect of placing both individual choice and social network ideas within the broader context of social scientific theory.

The following section uses the formal language, notation, and (generally) terms defined by Barlow and Proschan (1981). The same formal language as theirs will be used (which they developed for engineering reliability applications), but the *empirical meaning* of the formal language will be interpreted in an entirely different context, that of human decision-making groups. Interestingly, the "general systems theory" movement makes much ado about such analogies (see for example, Von Bertalanffy 1968; Casti 1989, pp. 9, 29), arguing that propositions can be made about certain types of general systems that will hold for specific cases of those systems. In this analysis, we find that the formal language and the derivations made within that formal language can be empirically interpreted in two very different specific contexts. This illustrates the point that the sought for general propositions of General Systems Theory are propositions about formal language systems. When two or more empirical phenomena can be interpreted according to the same formal language, then it is true that the derivations and predictions made within the

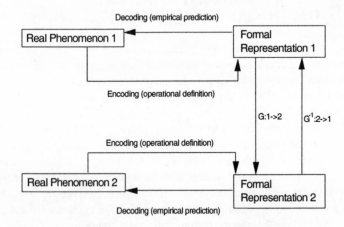

Figure 1. A model of system isomorphy

Notes: Real phenomenon 1 and Real phenomenon 2 are two different empirical objects of study, possibly
from different disciplines. Formal representation 1 is the formal theory of phenomenon 1. Formal
representation 2 is the formal theory of phenomenon 2. Encoding is the process of applying operational
definitions to the real phenomena, while decoding is the process of using the theory to make
predictions. G:1→2 indicates that there exists a function that maps representation 1 onto representation
2. G^{-1}:2→1 indicates that this function can be inverted, so that the two theories are isomorphic.

language apply to each specific phenomenon. This strategy, searching for a single
formal language that applies to disparate empirical phenomena, is known as
"systems isomorphy" or "systems analogy" (Von Bertalanffy 1968, p. 20). This
strategy departs from earlier social science attempts at using models from physical
science by focusing on the adequacy of the *formal language* in explaining each
phenomenon *separately*, and then considering the relationship between two formal
languages. This can be represented as given in Figure 1.

In this figure, the proper use of the strategy of system isomorphy is illustrated.
We begin with two real phenomena, each of which may be any object of empirical
study, from any discipline. Each of those real phenomena are encoded into a formal
representation or *theory*. Each theory generates model predictions, which are then
decoded via operational definitions into empirical predictions. The normal operation of a scientific discipline is circular, moving from real phenomenon to formal
representation and back, without influence from other models or phenomena.

There remains, however, the strategy of systems isomorphy, illustrated in the
figure by the arrows indicating a mapping from one formal theory to the other
formal theory. If a one-to-one and onto mapping exists from one formal language
to another, then we consider the two real phenomena to be isomorphic (Fararo 1989,
p. 353, note 13; Casti 1989, p. 9). This does *not* mean that the two real phenomena

are the same: Bacteria and human populations may both exhibit logistic growth, but they are clearly not the same. Likewise, it does *not* mean that the goodness of fit between real phenomenon and formal representation will be equal in both systems. What it *does* mean is this: The formal language (theory) developed in reference to one empirical phenomenon can be used as a theory with respect to the other empirical phenomenon. This is the claim of this paper: The formal language called coherent structure theory, developed in reference to engineering phenomena, is applicable to decision making and coalition formation in social networks.

A COHERENT STRUCTURE THEORY OF DECISION MAKING IN GROUPS AND SOCIAL NETWORKS

In coherent structure theory, components can occupy only one of two states: functioning or failed. In wholly analogous fashion, we can envision a decision maker being in one of two states: Voting "yes" on a measure, or voting "no." To indicate the state of the ith decision maker, we make this definition:

Definition 1. *Let the state of the ith decision maker be represented by a binary variable x_i taking on the value 1 or 0. Define a decision maker as voting yes if $x_i = 1$ and voting no if $x_i = 0$.*[4]

In addition to the state of the ith decision maker, we are also interested in the state of the group, organized in some fashion. Further, we will assume that the state of the group is entirely determined by the states of each member of the group. Thus, we make this definition:

Definition 2. *Let the state of the group be represented by a binary variable ϕ taking on the value 1 or 0. The state of the group is wholly determined by the state of each of n individual members of the group: $\phi = \phi(x)$, where $x = (x_1, \ldots, x_n)$. Thus, ϕ is a function mapping $(x_1, x_2, \ldots, x_n)$ into $\{0,1\}$. ϕ will be denoted a "structure function" or more simply, a "structure," that represents the way* individual votes *are pooled into a group vote.*

As an aid to representing structure functions, we will define two operators in this space of binary variables:

Series operator for two decision makers: $x_1 \sqcap x_2 \equiv x_1 x_2 = \min(x_1, x_2)$.

Parallel operator for two decision makers: $x_1 \sqcup x_2 \equiv 1-(1-x_1)(1-x_2) = \max(x_1, x_2)$.[5]

The intuitive meaning of these operators is not immediately apparent, and deserves some mention. The series operator is equivalent to a logical AND operation; that is, for $x_1 \sqcap x_2$ to equal one, x_1 AND x_2 must each equal one. Similarly, the parallel operator is equivalent to the logical OR operation; that is, for $x_1 \sqcup x_2$ to equal one x_1 OR x_2 must equal one.

Given these two definitions and operators, we can easily represent some common decision rules as functions mapping individual states (decisions) onto a collective decision. For example, unanimity rule requires that all persons in the group vote for a particular option. We can represent this by:

$$\phi(\mathbf{x}) = \sqcap_i x_i = \min(x_1, \ldots, x_n).$$

(In this case, all must vote "yes" for the group to choose the option. If any decision maker votes "no," the group will also vote "no.")

In contrast to unanimity rule, we can represent a voting structure where only one vote is necessary for the group to choose the option. This situation can be described as one in which if any one person votes for the option, the group will acquiesce (For example, we can imagine a group deciding to go out for coffee if any one individual is willing to pay for everyone.) This is represented by:

$$\phi(\mathbf{x}) = \sqcup_i x_i = \max(x_1, \ldots, x_n).$$

(In this case, only one decision maker must vote "yes" for the group to choose the option. All must vote "no" for the group to vote "no.")

Of course, somewhere between these two extremes lie the more common rules such as majority rule. These are referred to as k out of n structures (i.e., out of n decision makers, k must be decisive), and can be represented in several fashions. The simplest representation is the following:

$$\phi(\mathbf{x}) = I(\Sigma_i x_i \geq k);$$

where I is the indicator function taking on the value 1 if the argument is TRUE, and 0 if the argument is FALSE. This simply illustrates that at least one coalition of k out of n total decision makers must vote yes, and if any one decisive coalition exists the group structure will also vote "yes."

Suppose there existed an additional member of a group who took no part in the deliberation nor had a vote (e.g., a nonvoting observer). Though such an individual might have their own state of opinion, they would have no influence on the group's outcome. We will call such a member *irrelevant*.

Definition 3. *The ith decision maker is irrelevant to the structure ϕ if ϕ is constant in x_i; that is, if $\phi(1_i, \mathbf{x}) = \phi(0_i, \mathbf{x})$ for all combinations of $(\bullet_i, \mathbf{x})$. Otherwise the ith decision maker is relevant to the structure.*[6]

If the ith decision maker is irrelevant to the group's voting structure, this simply means that his or her vote does not make any difference to the group's vote. Using these three definitions, we now have the tools to define *monotonic* and *coherent* structures. There are two crucial purposes for these definitions: First, to eliminate from consideration any structure where a decision maker voting "yes" causes the group to vote "no" (an unreasonable case) and, second, to eliminate structures that contain irrelevant decision makers.

Definition 4. *A structure of decision makers is* monotonic *if and only if: (1) its structure function* ϕ *is nondecreasing in each argument.*[7] *A structure of decision makers is* coherent *if and only if: (1) it is monotonic, and (2) each decision maker is relevant to the structure.*

For purposes of illustrating the structures defined here, all coherent structures consisting of one, two, or three components are produced in Figure 2.

In this figure, consider each member as if he/she were a *switch* that could be turned *on* ("yes") or *off* ("no"). A yes-voting member connects the path on the left with the path on the right; a no-voting member breaks the path at that point. Moving from left to right, the group's decision is *yes* if a path exists from the farthest left of the diagram to the farthest right of the diagram. The group's decision is *no* if no such path can be drawn. The figure is a convenient graphical tool that clarifies the meaning of the coherent structure functions and makes the coalitions and decision-making structures visible.

If a decision-making group consists of only one person (structure one in Figure 2), obviously the only possible coherent structure is $\phi(x) = x$; the group's decision is that of its sole member. If a decision-making group consists of two persons, there are two possible coherent structures, one in which both must vote yes (structure two, $\phi(\mathbf{x}) = x_1x_2 = \min(x_1,x_2)$) and one in which only one must vote yes (structure three, $\phi(\mathbf{x}) = x_1 \sqcup x_2 = \max(x_1,x_2)$).

When a decision-making group has three members, the combinations increase rapidly. Moving from two to three members makes *coalition formation* possible. This is not a new discovery, having been considered in detail by Simmel (1950, pp. 118–177). However, what we see in this formal system is an exhaustive taxonomy of *all* possible decision-making structures that satisfy reasonable conditions. For example, obviously we have unanimity rule (structure four, $\phi(\mathbf{x}) = x_1x_2x_3$) and a one out of three rule (structure five, $\phi(\mathbf{x}) = x_1 \sqcup x_2 \sqcup x_3$), but we also find a coalition structure where decision makers two and three can together block passage of a vote but not pass the vote (structure six, $\phi(\mathbf{x}) = x_1(x_2 \sqcup x_3)$), thus indicating the guaranteed success of a 1-2 or a 1-3 coalition, but not a 2-3 coalition. We call this structure the "strong chair" structure since the chair can always block a 2-3 coalition—no motion will pass to which he or she objects.

One point of historical interest regarding structure six: Simmel (1950, p. 137) alludes to the political usefulness of Napoleon's Consulate of Three in this passage:

> Napoleon's Consulate of Three was decidedly more convenient for him than a group of two, for he had to win over only one colleague (which is very easy for the strongest among three) in order to dominate, in a perfectly legal form, the other, that is, actually, both other colleagues.

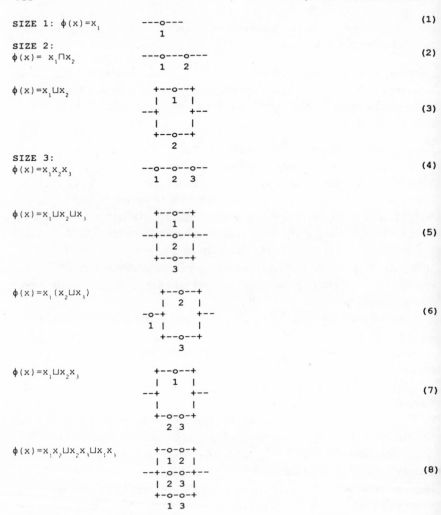

Figure 2. All coherent structures of size one, two, and three

Notes: Read from left to right; a member voting "yes" connects the path entering from the left with the path
exiting from the right; a member voting "no" breaks the path at that point; a group's decision is *yes* if
a path can be drawn from the far left to the far right, while the decision is *no* if no such path can be
drawn.

If person one (Napoleon) knows what he wants, or, more importantly, what he does
not want (i.e., $x_1 = 1$ or $x_1 = 0$), and can unilaterally block proposals by two and
three, then the structure is transformed from a two out of three structure (structure
eight) to one resembling structure six. Person one is clearly more able to control
the outcome of structure six than structure eight. This is important because the

interpretation of the structures matches the historical/theoretical interpretation of Simmel.

Considering the final two structures, we find a coalition structure where one alone can force passage of a vote over the "no" votes of two and three, and two and three together must form a coalition to force passage if one votes "no" (structure seven, $\phi(\mathbf{x}) = x_1 \sqcup (x_2x_3)$). Structure seven can be considered a "unilateral chair" structure, because the chair (person one) has the capability of passing a motion over the objections of person's two and three. Finally, we see a two out of three structure, where any two of the members can force passage (structure eight, $\phi(\mathbf{x}) = x_1x_2 \sqcup x_2x_3 \sqcup x_1x_3$). When we consider all coherent structures consisting of four members, the possible combinations (and their complexity) increase rapidly.

IMPLICATIONS OF COHERENT STRUCTURE DEFINITIONS

The first four definitions result in a substantial number of implications. Rather than follow the theorem-proof presentation style favored by mathematicians, we have chosen to include the major definitions and theorems in the Appendix. Instead, we wish to explore the *substantive* implications of the definitions and theorems of the theory. Particularly, and most importantly, we wish to discuss the implications coherent structure theory has for decision-making rules in groups, and show how coherent structures point the way to new types of group structures that have heretofore not been examined. Where necessary, we will indicate the theorem from which we draw our points.

Result 1: Upper and Lower Bounds on Group Competence

One important implication of coherent structure theory is that a group making pass-no-pass voting decisions have a group competence upper and lower bound. Theorem 3 (see Appendix) indicates that the performance of a group is bounded above by a parallel (1 out of *n*) decision rule and bounded below by a unanimity (*n* out of *n*) decision rule. This means that there is no decision rule that can be more likely to pass a measure than a *1 out of n* decision rule, and that there is no decision rule which can be less likely to pass a measure than an *n out of n* rule. The latter implication has also been found in empirical work described by Miller: Unanimity rule is less likely to reach a decision than any other rule. This bound is true both for coherent structures (where all decision makers are relevant) and for monotonic structures (where irrelevant decision makers are allowed).

Result 2: Implications of Minimal Paths and Importance of Structural Position

Definition 5 (see Appendix) describes a minimal path set in coherent structure theory; a minimal path set is a minimal "yes"-voting coalition necessary for passage of a vote. Any coalition larger than the minimal coalition and including it can also pass a vote, but any coalition smaller than the minimal coalition cannot. If one member of a minimal path set moves from decisive to indecisive, that coalition cannot pass a vote. However, this does not necessarily prevent some *other* coalition from passing a vote. For example, consider the structure $\phi(\mathbf{x}) = x_1 x_2 \sqcup x_3 x_4 x_5$. Obviously, the two minimal path sets are $\{1,2\}$ and $\{3,4,5\}$; if member number one votes "no," member number two cannot pass the vote. However, members three, four, and five can pass a vote if all vote "yes." In either set, if any one becomes a zero, the group also moves from one to zero (yes to no). Note the direct similarity between this definition of a minimal path and the definition of a minimal winning coalition from coalition formation research (Komorita and Kravitz 1983; Kravitz 1987).

The importance of determining minimal path structures can be seen by considering Napoleon's "Consulate of Three" in more detail. Durant and Durant (1975, pp. 159–160) describe the decision structure of the ruling consulate in this fashion. Immediately following the revolution, the three Provisional Consuls met to rebuild France:

> At their first sitting [Roger] Ducos proposed that the thirty-year-old general should take the chair. Bonaparte soothed [Emmanuel-Joseph] Sieyes by arranging that each of the three should preside in turn, and suggesting that Sieyes should take the lead in formulating a new constitution. The aging theorist retired to his study and left Napoleon (with Ducos complaisant) to issue decrees...

If Napoleon (person one) were equally powerful with person two (Emmanuel-Joseph Sieyes) and person three (Roger Ducos), a consulate of three operating under majority rule would obviously distribute power equally. But consider the case alluded to by Simmel, where Napoleon dominates the other two. There are a variety of possible ways to interpret this situation, but let us consider structure six from Figure 2, with structure function $\phi(\mathbf{x}) = x_1 (x_2 \sqcup x_3)$. What are the minimal paths of this structure? Clearly, $P_1 = \{1,2\}$ and $P_2 = \{1,3\}$. Thus, Napoleon, as chair and strong leader, can block any decision made by the 2-3 coalition. He cannot unilaterally pass a decision himself, but he *can* form *either* a 1-2 or 1-3 coalition as necessary. With Sieyes "out of the way," and Ducos complaisant (i.e., voting with Napoleon), then *Napoleon alone can pass or block any initiative he wants*. Obviously, such a situation has the surface appearance of a "democratic" triumvirate, but in fact is only slightly distant from a dictatorship. This illustrates the sociological truism that it is not *numbers* of persons but their *social structure* that determines outcomes.

Result 3:   The Existence and Importance of Subcommittees

It is also useful to consider situations in which more than one subgroup contribute their individual subgroup's decision to an overall collective decision. In coherent structure theory, such subgroups are called *modules*, reflecting their engineering origin. In the case of social decision-making structures, we will use the term "subcommittee," with the understanding that that which is called a subcommittee in social decision-making structures is called a module in engineering structures.

To understand the concepts being considered more intuitively, consider a situation where there exist three groups of three persons each. Assume that each group operates under a two out of three majority rule, but that each of the subgroups then passes their subgroup's decision to a central group decision which must be unanimous. Let the first three members be one subcommittee, the second three be the second subcommittee, and the last three be the third subcommittee. The structure function of such a group is:

$$\phi(\mathbf{x}) = (x_1x_2 \sqcup x_2x_3 \sqcup x_1x_3) \sqcap (x_4x_5 \sqcup x_5x_6 \sqcup x_4x_6) \sqcap (x_7x_8 \sqcup x_8x_9 \sqcup x_7x_9).$$

This structure function explicitly indicates that each subgroup is a two out of three structure while the group as a whole uses a three subcommittee unanimity rule. In addition and in contrast, consider a group with same first two subcommittees, but the third group consists of a single person. For example, imagine a departmental subcommittee voting on whether to make a job offer to a prospective new hire or not. They pass their decision on to the Dean of the college (x_7). It is impossible for the Dean to unilaterally pass a decision, but it is highly possible that the Dean can block a decision made by the subcommittees. This structure would be:

$$\phi(\mathbf{x}) = (x_1x_2 \sqcup x_2x_3 \sqcup x_1x_3) \sqcap (x_4x_5 \sqcup x_5x_6 \sqcup x_4x_6) \sqcap (x_7)$$

Both of these structures are illustrated graphically in Figure 3.

Figure 3 illustrates the presence of the three three-person subcommittees in the system. It further illustrates the coalitions necessary for passage of a vote. For example, if persons one, three, four, six, seven, and nine vote yes, then the group will vote yes; in each subgroup a two person coalition exists that carries the subgroup's vote, while all three subgroups form a unanimous coalition. Note that if persons one, two, three, four, five, six, and seven voted yes, but persons eight and nine did not, the group's decision will be *no*; though there exists a numerical majority in favor of the option, the structure of the voting situation guarantees that eight and nine alone have a veto power. This, and the strong Dean structure immediately following, illustrate another facet of social structure: There exist persons who, because of their structural position, have greater control over group outcomes. In the top structure of Figure 3, any two people within one subcommittee (not *any* two people) have the power to block a decision. In the bottom structure of Figure 3, any two people within one subcommittee have such power, but the strong Dean has a unilateral power to block, but not pass any action. Such people have been labeled "gatekeepers" by Lewin (1951) and Cartwright and Harary (1977).

```
                 Three equipowerful subcommittees

    +--o--o--+              +--o--o--+              +--o--o--+
    |  1  2  |              |  4  5  |              |  7  8  |
 ---+--o--o--+-------------+--o--o--+-------------+--o--o--+--
    |  2  3  |              |  5  6  |              |  8  9  |
    +--o--o--+              +--o--o--+              +--o--o--+
       1  3                    4  6                    7  9

    Subcommittee 1          Subcommittee 2          Subcommittee 3


            Two equipowerful subcommittees with a strong Dean
                        who can block decisions

    +--o--o--+              +--o--o--+
    |  1  2  |              |  4  5  |
 ---+--o--o--+-------------+--o--o--+--------------o-----
    |  2  3  |              |  5  6  |                 7
    +--o--o--+              +--o--o--+
       1  3                    4  6
    Subcommittee 1          Subcommittee 2      Subcommittee 3
```

Figure 3. Graphical illustration of a three subcommittee structure

Notes: Read from left to right; a yes-voting member connects the path entering from the left with the path
        exiting from the right; a no-voting member breaks the path at that point; a group's decision is *yes* if a
        path can be drawn from the far left to the far right, while the decision is *no* if no such path can be
        drawn.

Result 4: Measures of Structural Importance

In the Appendix, we provide a definition of structural importance suggested by Birnbaum (1969, Definition 8). In coalition terms, there are 2^{n-1} situations where the ith decision maker *could be* pivotal in determining whether the group passes a vote. The measure of importance indicates what proportion of the time the ith decision maker is *indeed* pivotal and determines the outcome of the group. The analogy between this measure of structural importance and other measures developed in decision theory contexts is important. Indeed, in 1965, Banzhaf developed a power index in coalition and game theory in which a person's power was equal to the number of times he/she was the "swing" vote in a coalition divided by the total number of possible voting combinations (see also Grofman and Owen 1982). This index, apparently developed independently, is equivalent to the Birnbaum index (Goldberg 1982, p. 7). Notably, Shapley and Shubik (1954) also developed a separate index based on different principles. This index has been contrasted with the Banzhaf index, but, apparently, never with the Birnbaum coherent structure version.

Result 5: Linking Structure to Individual Choice Theory

Up to now in this section we have considered only the *structural* properties of voting systems. That is, we have focused on whether or not a structure function ϕ

will equal zero or one when the individual choices x_i are known in advance. In behavioral applications, however, it is highly unlikely that individuals' choices at any one instant will be known with certainty. Instead, it is more likely that we can only assess each individual's probability of voting "yes" or "no." Thus, at this point we need to examine the probability models associated with coherent structures. Fortunately, coherent structure theory provides model development in that area, also. Theorem 6 results from the definition of a coherent structure as one which has no irrelevant decision makers. It says this: As each individual's probability of voting "yes" increases in a coherent decision-making structure, the group's probability of concluding "yes" also increases.

If passing a vote is defined as an arbitrary "correct" choice, then Theorem 6 implies that as individual competence increases, group competence also increases. (Recall that competence is defined as the probability of making a correct choice, for individual or group.) This is an important result: Given each individual's competence, but an *unknown* coherent structure, we can always calculate an upper and lower bound for the group's competence (Theorem 7 in the Appendix). These bounds need not be very narrow, but more precise bounds can also be calculated using pivotal decomposition (Theorem 1 in the Appendix) or modular (subcommittee) decomposition (see Barlow and Proschan 1981, pp. 34–44).

Result 6: Generating All Possible Coherent Structures of Any Size

Using the definitions and theorems presented up to now, the formalism of coherent structure theory not only allows us to generate various plausible decision rules, but it allows us to determine the whole universe of plausible decision rules. In the Appendix, we derive theorems which, collectively, give us the tools to generate all possible rules if we know what size group we want. Thus, these tools let us map out the group decision-making landscape in a way that was not fully possible before. Any group, of *any* size, that obeys the definition of coherent structures (Definition 4) in developing their decision rule, *must* use one of the rules that we can generate. The procedure is tedious, since the number of possibilities grows very quickly, but each step is quite simple. Illustrations of the technique for groups of size two and three are given in the Appendix.

LINKING COHERENT STRUCTURE THEORY AND GROUP DECISION THEORY

From this point, we use the concepts of coherent structure theory as foundations. We have already alluded to the decision and coalition theory implications of coherent structure theory. In this section we will proceed in three parts: First, we will develop the representation of common decision rules in coherent structure theory; second, we will show how common decision rules are modeled in traditional theory; third, we will show the relationship between coherent structure theory and traditional theory, and will show the advantages of coherent structure theory over traditional theory.

To begin to model common decision rules, let us consider a simple decision situation involving three persons of equal competence independently choosing between two options using a two out of three majority rule. In this system, we must specify the structure function that translates all possible combinations of individual votes into a group vote taking on the value one or 0. The structure function is:

$$\phi(\mathbf{X}) = X_1X_2 \sqcup X_2X_3 \sqcup X_1X_3$$

An immediate question that this structure generates is: What is the probability distribution of ϕ, the group's decision? That is, what are $P[\phi = 1]$ and $P[\phi = 0]$? To begin, consider the definition of ϕ. Note that the two conditions are mutually exclusive and exhaustive. If we know $P[\phi = 1]$, we can calculate $P[\phi = 0] = 1 - P[\phi = 1]$. We can solve this using the following enumeration:

$$P[\phi = 1]$$

$$= P[X_1,X_2 = 1 \text{ OR } X_2,X_3 = 1 \text{ OR } X_1,X_2,X_3 = 1]$$

$$= p \cdot p \cdot (1 - p) + (1 - p) \cdot p \cdot p + p \cdot (1 - p) \cdot p + p \cdot p \cdot p$$

$$= \sum_{k=2}^{3} \binom{3}{k} p^k (1 - p)^{3-k}$$

which, not coincidentally, is the same answer as that given by Condorcet's theorem. From Condorcet's theorem we know that if $p > .5$, $P[\phi = 1] > .5$, and if $p < .5$, $P[\phi = 1] < .5$. More importantly, from Theorem 7 (see Appendix), we know that, under independent choices, the group's competence *cannot* be greater than $1-(1-p)^3$ or less than p^3. For example, if the competence of each individual was .75, the group's competence, for any possible coherent structure, must be greater than .4219 and less than .9844. In this situation, under 2 out of 3 majority rule, the group's probability of passing a vote is .8438.

WEIGHTED MAJORITY DECISION RULES

One important test of the theoretical usefulness of coherent structure theory is its ability to model common decision rules. In the introductory section, we discussed several: Weighted and unweighted majority rule(s), unanimity rule, veto rule(s), and dictatorship are among them. The important question is this: Do there exist coherent structures that are capable of representing these decision rules?

To answer this question, we need to find the common factors involved in these rules. One function of high generality which encompasses all of the above is the following decision rule, with parameters w_i ($i = 1, 2, \ldots, n$) and k:

$$R(x: w,k) = I\left[\sum_{i=1}^{n} w_i x_i \geq k \right]$$

where $R(x{:}w,k)$ = the outcome of the group's vote, a function mapping the group votes x onto $\{1,0\}$ with parameters w and k; I = the "indicator" function, taking on the value 1 when the argument is TRUE, and 0 when the argument is FALSE; w_i = the weight given the vote of the ith decision maker, ($w_i \in (0, +\infty)$); k = the total weight necessary to choose option one ($k \leq \Sigma_i w_i$).

The parameters w_i correspond to Miller's (1989) *distribution of power* dimension, and the parameter k corresponds to his *strictness* dimension. That is, w_i determine by their values the relative power of each individual in terms of their influence on the group vote. K, on the other hand, determines the amount of weighted votes needed to force the group's vote in one direction or the other. By appropriately specifying these parameters, the following decision rules given in Table 1 are easily modeled. As can be seen, in each case the common language

Table 1. Some Common Decision Rules

Unweighted greater than 50% majority rule:

$w_i = 1$, for all i; $k = [(n + 1)/2]$ (*i.e., the smallest integer greater than* $(n + 1)/2$);

$$R(x: w, k) = I\left[\sum_{i=1}^{n} x_i \geq k \right]$$

Weighted greater than 50% majority rule:

$$k = \left[\left(\sum_{i=1}^{n} w_i + 1 \right)/2 \right] \ (\textit{i.e., the smallest integer greater than} \left(\sum_{i=1}^{n} w_i + 1 \right)/2)$$

$$R(x: w, k) = I\left[\sum_{i=1}^{n} w_i x_i \geq k \right]$$

Unanimity rule:

$w_i = 1$, for all i; $k = n$;

$$R(x: w, k) = I\left[\sum_{i=1}^{n} x_i = n \right]$$

Veto rule:

$w_n \neq 0$; $k > \sum_{i \neq n} w_i$;

$$R(x: w, k) = I\left[\sum_{i \neq n} w_i x_i + w_n x_n \geq k \right]$$

Dictatorship:

$w_n = 1$; $k = 1$

$R(x: w, k) = I\ [x_n = 1]$

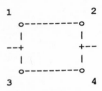

Figure 4. Graphical representation of a non-weighted-majority structure

Notes: Read from left to right; a yes-voting member connects the path entering from the left with the path exiting from the right; a no-voting member breaks the path at that point; a group's decision is *yes* if a path can be drawn from the far left to the far right, while the decision is *no* if no such path can be drawn. Notice that it is impossible to choose votes such that, for example, persons one and two can pass a vote while persons two and four cannot, as required by the coherent structure.

description of the decision rule is translated into a statement about the votes x_i and parameters w_i and k. Given this form for various common decision rules, the question is this: *Do weighted majority rules encompass all possible reasonable decision rules for groups?*

Since we have already proposed coherent structures as a set of reasonable definitions for group decision-making rules, the question becomes transformed to: Can we find a coherent structure for each R such that $\phi(x) = R(x{:}w,k)$ for all possible combinations of x? Theorem 8 indicates that we can, and shows us how—so that any weighted majority rule can be represented as a coherent structure.

In order to complete the original question (do weighted majority rules encompass all possible reasonable decision rules for groups?), we also reverse the question: Is the converse true? Can all coherent structures be represented by a weighted decision rule R?

An immediate answer to this question lies in finding an example of a coherent structure that cannot be represented in the form of R. Such structures do exist; in fact, there are *numerous* coherent (and reasonable) group decision-making rules that *cannot* be represented as a weighted majority rule. This is a structure in which the "yes"-voting coalitions $\{1,2\}$, $\{3,4\}$ suffice to pass a vote. This structure is presented graphically in Figure 4. Note that no combination of votes can be assigned to these four actors such that, for example, persons one and two can pass a vote but persons one and four cannot.

This example illustrates that while all combinations of common rules (R) can be represented by coherent structures, the converse is not true. Thus, we have demonstrated here that coherent structures are a more general class of group decision-making rules than those representable by R. This is not a trivial conclusion; several authors in coalition theory limit their focus to decision rules that take the form of a weighted majority (e.g., Mann and Shapley 1964). We conjecture that this limitation is a source of difficulty for group decision-making theory, because there are plausible group decision structures that are *not* representable by a weighted majority rule. Thus, the larger class of coherent structures may prove a fruitful avenue of exploration than the smaller class of weighted majority rules, especially for empirical investigations of decision-making groups.

Now we come to the final question: What classes of coherent structures *cannot* be represented by a weighted majority decision rule? That is, how do we know we have a coherent structure collective decision-making rule that we cannot reduce to a weighted majority rule? The results in the Appendix leading up to now provide sufficient information to generate the following theorem, which we shall explore in some detail.

Theorem 11: Let P_j for some $j = 1,2, \ldots, r$ be a set of minimum paths for a coherent structure (C, ϕ). Then (C, ϕ) is *not* representable by a weighted majority rule of the form R if: (1) there exist at least two min-path sets, $\{x\}_i$ and $\{x\}_k$, $i \neq k$, each of which has two or more components, and (2) at least two elements of each min-path set are disjoint.

This result indicates that if there are two or more minimal coalitions with 2 or more decision makers each, then ϕ cannot be represented by R. For example, the simple case illustrating the rule, given in Figure 4, is the structure function with minimal

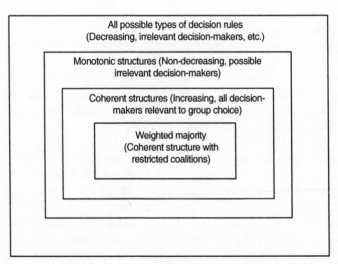

Figure 5. The hierarchy of decision rules

Notes: Each set of decision rules encompasses the set within it. Thus, weighted majority rules are a subset of coherent structures due to limitations on the types of structures that can be formed. Coherent structures are a subset of monotonic structures because all decision-makers must be relevant to the structure. Monotonic structures are a subset of all possible structures since decision rules can be imagined which are decreasing, weights are negative, all decision makers are irrelevant, and so on.

coalitions {1,2},{3,4}. Further, it is not difficult at all to write minimal coalitions that cannot be represented as weighted majority rules,[8] although these meet the definitions of a reasonable decision structure.

To summarize the implications of these results: We have shown that there exists a class of group decision-making rules that contain irrelevant decision makers and negative and positive weights on each decision maker. Within this class, there are the rules for monotonic structures which are nondecreasing (i.e., nonnegative weights), but have irrelevant decision makers. Within this class, there are rules which are increasing and have all relevant decision makers. Finally, the most restricted class is weighted majority rules, with positive weights and specially restricted coalitions. Thus, the different types of decision rules are arranged in a hierarchy of classes, as illustrated in Figure 5.

CONCLUSIONS

The conclusions to this paper can be summarized in four points:

- First, we have shown that the mathematical theory of coherent structures provides a formal structure that organizes and summarizes some historical and experimental results in coalition formation and group decision making.
- Second, we have shown that it is possible to model many forms of voting and interaction rules with coherent structures, and certain coherent structures are equivalent to common decision rules.
- Third, we have demonstrated that there exist plausible decision rules not representable by a simple weighted voting scheme.
- Fourth, for each concept discussed by the developers of coherent structures (e.g., minimal paths, modules, monotonic structures, etc.), there is a parallel concept in the theory of coalitions (e.g., minimal coalitions, subcommittees, irrelevant group members, etc.).

Originally designed for components of mechanical and electronic systems, we have shown that it is potentially interpretable as a theory of group decision making and coalition formation. Furthermore, it is a more general theory of decision making because there exist certain decision and coalition situations that can be represented as coherent structures that *cannot* be represented as weighted majority rules. Since virtually all coalition and group decision rules studies make use of one form or another of weighted majority rule, this research clearly points the way to *new, unexplored* structures that can be analyzed both theoretically, experimentally, and historically. The theoretical and empirical implications of these structures may explain much that is troublesome about coalition and decision-making behavior.

Future work in this area should address several questions and theoretical issues. First, in this paper we assumed that coalition structures were predetermined, known to the experimenter, and fixed. Each of these aspects can be examined. Much of coalition theory (e.g., Caplow 1956; Gamson 1964; Komorita and Kravitz 1983) focuses on predicting the coalition structure that will occur given a certain set of initial resources distributed unequally. A potentially fruitful avenue of research is to link these theories on the "front end" of the current work, so that the ultimate decision structure of a group can be predicted only from knowledge of its starting resource allocation. Further, we have assumed that the experimenter knows without error what coherent structure is operative in any particular group. An empirical approach to group decisions could focus on determining the distribution of possible coalition structures from observing the behavior of individuals. For example, Kushner and DeMaio (1977) developed a method of determining crucial actors in a voting body using directed graphs. Approaches like these can be used to determine what types of structures best fit the obtained data. Finally, like many structure theories, this one assumes that structures are fixed. A relatively untouched area of research is that of *structures* that change over time. Structures obviously change, as an historical analysis of coalition patterns (de Swaan 1973) indicates. The change in structures can be simulated in the laboratory through some arbitrary point system in an experimental game, but can these changes be identified and predicted in real world interaction? A working solution to this problem would move social network research ahead. A final agenda for future research is that agenda that applies to all *scientific* theories: That it be tested against data to determine its adequacy.

APPENDIX
DEFINITIONS, THEOREMS AND PROOFS OF THEOREMS IN COHERENT STRUCTURE THEORY AND COLLECTIVE DECISION RULES

Definition 1. *Let the state of the ith decision maker be represented by a binary variable x_i taking on the value 1 or 0. Define a decision maker as voting yes if x_i = 1 and voting no if x_i = 0.*

Definition 2. *Let the state of the group be represented by a binary variable ϕ taking on the value 1 or 0. The state of the group is wholly determined by the state of each of n individual members of the group: $\phi = \phi(x)$, where $\mathbf{x} = (x_1, \ldots, x_n)$. Thus, ϕ is a function mapping $(x_1, x_2, \ldots, x_n)$ into $\{0,1\}$. ϕ will be denoted a "structure function" or more simply, a "structure."*

Definition 3. *The ith decision maker is irrelevant to the structure ϕ if ϕ is constant in x_i; that is, if $\phi(1_i, \mathbf{x}) = \phi(0_i, \mathbf{x})$ for all combinations of $(\bullet_i, \mathbf{x})$. Otherwise the ith decision maker is relevant to the structure.*[*]

Definition 4. *A structure of decision makers is monotonic if and only if: (1) its structure function ϕ is nondecreasing in each argument.*[**] *A structure of decision* *makers is* coherent *if and only if: (1) it is monotonic, and; (2) each decision maker is relevant to the structure.*

> **Theorem 1:** **(Pivotal decomposition).** The following identity holds for any structure function ϕ of order n:

$$\phi\mathbf{x} = x_i\phi(1_i,\mathbf{x}) + (1 - x_i)\phi(0_i,\mathbf{x})$$

Proof. See Barlow and Proschan (1981, p. 5).

> **Theorem 2:** A coherent structure function has at least one relevant component if and only if $\phi(\mathbf{0}) = 0$ and $\phi(\mathbf{1})=1$.

Proof. (proof for "if"):
1. Let $\phi(\mathbf{0}) = 0$ and $\phi(\mathbf{1}) = 1$.

BY CONTRADICTION:

2. Let $\phi(1_i,\mathbf{x}) = \phi(0_i,\mathbf{x})$ for all i (all components are irrelevant).
3. Then $\phi(\mathbf{1}) = \phi(\mathbf{0})$, which contradicts assumption in step 1.
4. Therefore, there exists an ith component such that $\phi(1_i,\mathbf{x}) \neq \phi(0_i,\mathbf{x})$.
5. By definition, component i is relevant.

(proof for "only if"):

6. Let ϕ be increasing in each argument.
7. Let ϕ have at least one relevant component i.
8. By definition, i relevant implies $\phi(1_i,\mathbf{x}) \neq \phi(0_i,\mathbf{x})$.
9. Since ϕ is increasing in each argument and $\phi(\mathbf{x}) \in \{0,1\}$, $\phi(0_i,\mathbf{x}) \leq 1$ and $\phi(1_i,\mathbf{x}) \geq 0$.
10. Since $\phi(1_i,\mathbf{x}) \neq \phi(0_i,\mathbf{x})$, and both are $\in \{0,1\}$, either

$$(5.1) \quad \phi(1_i,\mathbf{x}) = 0 \text{ and } \phi(0_i,\mathbf{x}) = 1,$$

which is not increasing and hence contradicts (6), or

$$(5.2) \quad \phi(1_i,\mathbf{x}) = 1 \text{ and } \phi(0_i,\mathbf{x}) = 0.$$

11. Since i is arbitrarily chosen,

$$\phi(1_i,\mathbf{x}) = 1 \text{ implies that } \phi(\mathbf{1}) = 1 \text{ and}$$

$$\phi(0_i,\mathbf{x}) = 0 \text{ implies that } \phi(\mathbf{0}) = 0.$$

> **Theorem 3:** Let ϕ be a coherent structure of n components. Then the following inequalities hold:

$$\prod_{i=1}^{n} {}_i x_i \leq \phi(x) \leq \bigsqcup_{i=1}^{n} {}_i x_i$$

Proof. See Barlow and Proscan (1981, p. 6).

Lemma 1. Let $\phi^R(0,x)$ and $\phi^R(1,x)$ be structures with all occurrences of irrelevant decision makers replaced by 0 or 1, respectively. For $\phi(x)$ a monotonic structure, $\phi^R(0,x)$ and $\phi^R(1,x)$ are coherent structures.

Proof. In the monotonic structure function ϕ replace all occurrences of irrelevant decision makers by 1; then the remaining structure function is monotonic (by assumption), and all remaining decision makers are relevant. Therefore the structure function $\phi^R(1,x)$ is coherent by the definition of coherence. The same reasoning applies to $\phi^R(0,x)$.

Theorem 4: Let ϕ be a monotonic structure with $1, 2, \ldots, r$ relevant decision makers and $r+1, r+2, \ldots, n$ irrelevant decision makers. Then:

1. $\displaystyle\prod_{i=1}^{r} x_i \leq \phi(x) \leq \bigsqcup_{}^{r} x_i$, and

2. $\phi(0) = 0$ and $\phi(1) = 1$.

Proof. If the structure is coherent ($r = n$; all decision makers are relevant), these results both hold, by the theorems above. Thus, we will deal only with the case where there exist irrelevant decision makers. Without loss of generality, let the first r decision makers be relevant and the $r+1, \ldots, n$ decision makers be irrelevant. Recall that $\phi(1_j,x) = \phi(0_j,x)$ for all x defines irrelevance for the jth decision maker.

1. Let $\phi^R(1,x)$ be the structure $\phi(x)$ with all irrelevant decision makers equal to 1, and let $\phi^R(0,x)$ be the structure $\phi(x)$ with all irrelevant decision makers equal to 0.
2. Since $\phi(0_{r+1}, x) = (1_{r+1}, x)$ for *all* x, then $\phi^R(0,x) = \phi^R(1,x)$ for all x.
3. Since $\phi^R(0,x) = \phi^R(1,x)$ for all x, $\phi(x) = \phi^R(0,x) = \phi^R(1,x)$ for all x.
4. We know that $\prod_{i=1}^{n} x_i \leq \phi^R(1,x) = \phi(x) = \phi^R(0,x) \leq \bigsqcup_{i=1}^{n} x_i$ (by Lemma 1).
5. Taking the left hand inequality, $\prod_{i=1}^{n} x_i \leq \phi^R(1,x) \Rightarrow \prod_{i=r+1}^{n} 1 \prod_{i=1}^{r} x_i \leq \phi^R(1,x)$, since we know that all irrelevant components equal 1. But $\prod_{i=r+1}^{n} 1 = 1$, so $\prod_{i=1}^{r} x_i \leq \phi^R(1,x)$.
6. Taking the right hand inequality, $\phi^R(0,x) \leq (\bigsqcup_{i=1}^{r} x_i) \sqcup (\bigsqcup_{i=r+1}^{n} 0)$, since we know that all irrelevant components equal 0. But $\bigsqcup_{i=r+1}^{n} 0 = 0$, so $\phi^R(0,x) \leq \bigsqcup_{i=1}^{r} x_i$.
7. Therefore, $\prod_{i=1}^{r} x_i \leq \phi^R(1,x) = \phi(x) = \phi^R(0,x) \leq \prod_{i=1}^{r} x_i$. This proves part 1.
8. To prove part 2, $\prod_{i=1}^{r} 0_i \leq 0 \leq \phi(x) \leq 0 \leq \bigsqcup_{i=1}^{r} 0_i$ as can be seen above.

9. Obviously, $0 \leq \phi(\mathbf{0}) \leq 0$ implies $\phi(\mathbf{0}) = 0$. Likewise, $\sqcap_{i=1}^{r} 1_i \leq 1 \leq \phi(x) \leq$
    $1 \leq \sqcup_{i=1}^{r} 1_i$ implies $\phi(\mathbf{1}) = 1$.   **Q.E.D.**

Definition 5.   *Let* $\mathbf{x}$ *indicate the states of the set of decision makers* $C = \{1, \ldots, n\}$.

(5.1)   Define $C_0(\mathbf{x}) = \{i | x_i = 0\}$ and $C_1(\mathbf{x}) = \{i | x_i = 1\}$.

(5.2)   A path vector is a vector $\mathbf{x}$ such that $\phi(\mathbf{x}) = 1$. The corresponding path set
        is $C_1(\mathbf{x})$.

(5.3)   A minimal path vector is a path vector $\mathbf{x}$ such that $\mathbf{y} < \mathbf{x} \Rightarrow \phi(\mathbf{y}) = 0$. The
        corresponding minimal path set is $C_1(\mathbf{x})$.***

Definition 6.   *Suppose that the state* X_i *of the ith decision maker is a binary*
random variable. Then $P[X_i = 1] = E(X_i) = p_i$. *Since* X_i *is a random variable,* $P[X_i$
$= 0] = 1 - P[X_i = 1] = 1 - p_i$. *Since each* X_i *is a random variable,* $\phi(\mathbf{X})$ *is also a*
random variable, and we will denote $P[\phi(\mathbf{X}) = 1] = E(\phi(\mathbf{X})) = h$. *If all the*
individuals make statistically independent judgements (that is, if all the random
variables $\mathbf{X}$ *are independent), we will represent the structural expected value* h *as*
a function of the individual expected values $\mathbf{p}$; *thus,* $h = h(\mathbf{p})$

Theorem 5:   Let $\mathbf{X}$ be a set of n statistically independent decisions. Then:

$$h(\mathbf{p}) = p_i \cdot h(1_i, \mathbf{p}) + (1 - p_i) \cdot h(0_i, \mathbf{p}) \text{ for } \mathbf{I} = 1, 2, \ldots, n.$$

Proof.   See Barlow and Proschan (1981, p. 21).

Theorem 6:   Let $h(\mathbf{p})$ be the expected value of a coherent structure ϕ. Then
                 $h(\mathbf{p})$ is strictly increasing in each p_i for

$$0 < < \mathbf{p} < < 1.****$$

Proof.   See Barlow and Proschan (1981, p. xx).

Theorem 7:   Let ϕ be a coherent structure of independent individuals with
                 individual competencies p_i. Then

$$\prod_{i=1}^{n} p_i \leq P[\phi(\mathbf{X}) = 1] \leq \bigsqcup_{i=1}^{n} p_i$$

Proof.   See Barlow and Proschan (1981, p. 34).

Theorem 8:   Let $R(x:w,k) = I\left[\sum_{i=1}^{n} w_i x_i \geq k\right]$ be not identically equal to zero
                 or one for some $n > 0$, $w_i \in (0, +\infty)$, $k \in (0, \infty)$, binary x_i. Let this
                 structure be increasing in each x_i. Then there exists a coherent

structure function ϕ such that $\phi(\mathbf{x}) = R(\mathbf{x};\mathbf{w},k)$ for all possible combinations of $\mathbf{x}$.

Proof (constructive).

1. Determine all of the $l = 1, 2, \ldots, p$ combinations $\mathbf{x}$ such that $\Sigma_i w_i x_i \geq k$ and $\mathbf{y} < \mathbf{x}$ implies $R = 0$. Denote the set $l = \{x_i | x_i = 1\}$ for $l = 1, \ldots, p$.
2. Let $\rho_l = \sqcap_{x_j \in l} x_j$.
3. Let $\phi(x) = \sqcup_{l=1}^{p} \rho_l = \sqcup_{l=1}^{p} \sqcap_{x_j \in l} x_j$.
4. Clearly, $R = 1$ if the lth set of votes all equal one, regardless of the other votes in the group. Likewise, $\mathbf{y} < \mathbf{x}$ implies $R = 0$. But $\rho_l = 1$ implies $\phi(\mathbf{x}) = 1$ if the lth set of votes all equal 1 and $\rho_l = 0$ implies $\phi(\mathbf{x}) = 0$ if $\mathbf{y} < \mathbf{x}$.
5. Therefore, $\phi(x) = \sqcup_{l=1}^{p} \rho_l = \sqcup_{l=1}^{p} \sqcap_{x \in l} x_j = R(\mathbf{x};\mathbf{w},k)$ for all possible combinations of $\mathbf{x}$. **Q.E.D.**

Definition 7. *The* structural importance *of the ith component for structure ϕ is given by:*

$$I_\phi(i) = \frac{1}{2^{n-1}} \sum_{[x|x_i=1]} [\phi(1_i,x) - \phi(0_i,x)]$$

where n is the size of the structure and the summation is taken over all vectors x such that the ith component is equal to 1.

Theorem 9: Let $P_1, \ldots, P_r$ be nonempty sets of components no one of which is a subset of any other. Let $C = \cup_{j=1}^{r} P_j$. Finally, let $\phi(\mathbf{x}) = 1$ if and only if $C_1(x) \supseteq P_j$ for some $j = 1, 2, \ldots, r$; otherwise $\phi(\mathbf{x}) = 0$. Then (C, ϕ) is a coherent structure with minimal path sets $P_1, \ldots, P_r$.

Proof.
1. $\phi(\mathbf{x}) = 1$ iff $P_j \subseteq C_l(\mathbf{x})$; thus, $\phi(\mathbf{x}) = 1$ iff the components in set P_j all equal 1. Otherwise, $\phi(\mathbf{x}) = 0$.
2. By the definition of a minimal path set, P_j is a minimal path set.
3. Since $P_1, \ldots, P_r$ are nonempty sets and no one of which is a subset of any other, it is not the case that $P_i \cup P_j = P_i$ for any i, j. Therefore, $C = \cup_{j=1}^{r} P_j$ is a set of components such that each component is a member of one or more sets.
4. Since a coherent structure can be represented as $\phi(x) = \sqcup_{j=1}^{r} \rho_j(x)$, and all relevant components are elements of some min-path set, then (C, ϕ) is a coherent structure with minimal path sets $P_1, \ldots, P_r$.

Definition 8. *Let $A = \{x_1, x_2, \ldots, x_n\}$ be a set of components, and let $\{B_i\}$ $(i = 1, 2, \ldots)$, be a set of nonempty subsets of A. $\{B_i\}$ $(i = 1, 2, \ldots)$ is called a* partition *of A if:*

1. $\cup_i B_i = A$; and
2. For any B_i, B_j, either $B_i = B_j$ or $B_i \cap B_j = \varnothing$.

Definition 9. *Let $A = \{x_1, x_2, \ldots, x_n\}$ be a set of components. Then the* power set *of A is the set of all subsets of A. If the set A is finite, say with n elements, then the power set of S has 2^n elements.*

> **Theorem 10:** Let $\{x_1, \ldots, x_n\}$ be a set of components. Then all coherent structures of order n may be constructed using the following procedure:

1. Find all partitions of $\{x_1, \ldots, x_n\}$. Label each element of each partition a_{1j_1}, $a_{2j_2}, \ldots, a_{kj_k}$ where a_{ij_i} is the ith partition, consisting of j_i subsets a_{i1}, a_{i2}, $\ldots, a_{ij_i}$.
2. Let $\mathbf{A} = \{\ \{a_{1.}\}, \{a_{2.}\}, \ldots, \{a_{k.}\}\ \}$ be the set of all partitions, where $\{a_{i.}\} = \cup_{j=1}^{r} a_{ij}$ (union of all the elements of the ith partition).
3. Calculate the power set of A: Label each element of the power set b_{1m_1}, $b_{2m_2}, \ldots, b_{nm_n}$, where $n = 2^k$, and b_{im_i} is the ith subset consisting of m_i elements $b_{i1}, b_{i2}, \ldots, b_{im_i}$.
4. Let $D_i = \cup_{m=1}^{m_i} b_{im}$.
5. The set D_i $(i = 1, 2, \ldots, n)$ is the set of min-paths of all coherent structures of order n. This includes permutations of components having the same structure.

Proof.

1. By the definition of a partition, the set $a_i = \cup_{j=1}^{j_i} a_{ij}$ includes all elements of the set $\{x_1, \ldots, x_n\}$.
2. By the definition of a power set, the set b_{1m_1}, $b_{2m_2}, \ldots, b_{nm_n}$ contains all possible combinations of partitions.
3. Since $A \subseteq B$ implies that $A \cup B = B$, then the union of all possible combinations of partitions eliminates all sets b_{ij} that are subsets of some other set b_{ik}.
4. Therefore, $D_i = \cup_{m=1}^{m_i} b_{im}$ contains all elements of $\{x_1, \ldots, x_n\}$, and no set is a subset of any other.
5. By Theorem 9, let $\phi(\mathbf{x})$ equal a parallel structure with each module (minimal path) equal to D_i. Then (C, ϕ) is a coherent structure.

Computational Corollary. Let $\{x_1, \ldots, x_n\}$ be a set of components. In step 3 in the previous theorem, the partitions consisting of $\{x_1 \ldots x_n\}$ (the whole set) and $\{x_1\}, \ldots, \{x_n\}$ (each individual element) need not be used to calculate the power set. These two partitions are included as min-paths, however.

Proof.

1. Since $A \subseteq B$ implies that $A \cup B = B$, then the union of all possible combinations of partitions eliminates all sets b_{ij} that are subsets of some other set b_{ik}.
2. But, the set $\{x_1, \ldots, x_n\}$ is a superset of any other partition, thus eliminating all other subsets with which it is unioned in a power set.
3. Likewise, the elements of the set $\{x_1\}, \ldots, \{x_n\}$ will always be subsets of any other partition with which they are unioned in a power set.
4. Since both are possible minimal paths, both are coherent structures. Thus, both must be included in the set of coherent structures.

To illustrate the operation of this theorem, first consider the set consisting of two components $\{x_1, x_2\}$. We wish to find all coherent structures that can be formed with these two components. Following the steps outlined above:

1. First we find all partitions of this set. These include $\{x_1 x_2\}$, and $\{x_1\}\{x_2\}$.
2. Let $A = \{ \{x_1x_2\}, \{x_1\} \cup \{x_2\} \}$.
3. By the computational corollary, we need not calculate the power set of A.
4. Let $D_1 = \{x_1 x_2\}$, and $D_2 = \{\{x_1\},\{x_2\}\}$.
5. These are the two possible coherent structures of order 2.

As a second example, consider the set consisting of three components, $\{x_1, x_2, x_3\}$. Again we follow the steps outlined above:

1. Find all partitions of this set. These are $\{x_1x_2x_3\}$, $\{x_1x_2\} \{x_3\}$, $\{x_1x_3\} \{x_2\}$, $\{x_2x_3\} \{x_1\}$, and $\{x_1\} \{x_2\} \{x_3\}$.
2. Let $A = \{ \{x_1x_2x_3\}, \{x_1x_2\} \cup \{x_3\}, \{x_1x_3\} \cup \{x_2\}, \{x_2x_3\} \cup \{x_1\}, \{x_1\} \cup \{x_2\} \cup \{x_3\} \}$.
3. By the computational corollary, reduce this set A to $\{ \{x_1x_2\} \cup \{x_3\}, \{x_1x_3\} \cup \{x_2\}, \{x_2x_3\} \cup \{x_1\} \}$. The power set of this reduced A is:

   $\{ \{x_1x_2\} \cup \{x_3\}, \{x_1x_3\} \cup \{x_2\}, \{x_2x_3\} \cup \{x_1\} \}$;
   $\{ \{x_1x_2\} \cup \{x_3\}, \{x_1x_3\} \cup \{x_2\} \}$;
   $\{ \{x_1x_3\} \cup \{x_2\}, \{x_2x_3\} \cup \{x_1\} \}$;
   $\{ \{x_1x_2\} \cup \{x_3\}, \{x_2x_3\} \cup \{x_1\} \}$;
   $\{ \{x_1x_2\} \cup \{x_3\} \}$;
   $\{ \{x_1x_3\} \cup \{x_2\} \}$; and
   $\{ \{x_2x_3\} \cup \{x_1\} \}$.

4. Let $D_1 = \{x_1x_2\} \cup \{x_3\} \cup \{x_1x_3\} \cup \{x_2\} \cup \{x_2x_3\} \cup \{x_1\} = \{x_1x_2\},\{x_1x_3\},\{x_2x_3\}\}$;
   let $D_2 = \{x_1x_2\} \cup \{x_3\} \cup \{x_1x_3\} \cup \{x_2\} = \{x_1x_2\},\{x_1x_3\}\}$;
   let $D_3 = \{x_1x_3\} \cup \{x_2\} \cup \{x_2x_3\} \cup \{x_1\} = \{x_1x_3\},\{x_2x_3\}\}$;

let $D_4 = \{x_1x_2\} \cup \{x_3\} \cup \{x_2x_3\} \cup \{x_1\} = \{x_1x_2\},\{x_2x_3\}$;

let $D_5 = \{x_1x_2\} \cup \{x_3\} = \{x_1x_2\},\{x_3\}$;

let $D_6 = \{x_2x_3\} \cup \{x_1\} = \{x_2x_3\},\{x_1\}$;

let $D_7 = \{x_1x_3\} \cup \{x_2\} = \{x_1x_3\},\{x_2\}$; and, returning the two discarded sets,

let $D_8 = \{x_1x_2x_3\}$; and

let $D_9 = \{x_1\},\{x_2\},\{x_3\}$.

5. The reader can verify that these sets are equivalent to the diagram of all possible coherent structures, except that all permutations of elements are included in these sets. D_8 corresponds to the unanimity rule structure, while D_9 corresponds to the one out of three structure. D_1 is the two out of three structure. D_2, D_3, and D_4 are the strong chair structures, with different persons occupying the strong chair position. D_5, D_6, and D_7 are unilateral chair structures, again with different persons occupying the unilateral chair position. Obviously, this procedure can be applied to any set of components. Thus, this theorem gives an algorithm for constructing all possible coherent structures of any size. Computing these by hand would quickly become impossible, or, at best, terribly tedious. However, the steps outlined are *merely* tedious, and can be implemented with a fairly simple computer algorithm.

Theorem 11: Let P_j for some $j = 1, 2, \ldots, r$ be a set of minimum paths for a coherent structure (C, ϕ). Then (C, ϕ) is not representable by a weighted majority rule of the form R if:

1. there exist at least two min-path sets, $\{x\}_i$ and $\{x\}_k$, $i \neq k$, each of which has two or more components, and
2. at least two elements of each min-path set are disjoint.

Proof.

1. Assume that there exists 2 disjoint min-paths.
2. Assume that the structure is representable by a weighted majority rule of the form R.

Then, by contradiction,

Without loss of generality, let the first two components of each of the first and second min-path sets be the disjoint components. Let the first two components of the first min-path set have weights w_1 and w_2. Let the first two components of the second min-path set have weights w_3 and w_4.

By Assumption (2),

$$(3.1) \quad w_1 + w_2 \geq k$$

and

$$(3.2) \quad w_3 + w_4 \geq k.$$

Adding both sides of each inequality in (3) together results in:

$$(4.1) \quad w_1 + w_2 + w_3 + w_4 \geq 2k.$$

However, also by Assumption (2), rearrange weights:

$$(5.1) \quad w_1 + w_3 < k$$

and

$$(5.2) \quad w_2 + w_4 < k$$

(since these components are not members of min-paths their weights are not equal to or greater than the parameter k).

Adding both sides of each inequality in (5) together results in:

$$(4.1) \quad w_1 + w_2 + w_3 + w_4 < 2k.$$

(6) contradicts (4). Since, by Assumption (1), 2 min-paths exist (and this fits the definition of a coherent structure), it must be the case that Assumption (2) is false. Therefore, the structure *cannot* be represented by a weighted majority rule R. **Q.E.D.**

APPENDIX NOTES

* This notation indicates the following:

$\phi(1_i, \mathbf{x}) \equiv (x_1, \ldots, x_{i-1}, 1, x_{i+1}, \ldots, x_n)$, which means that the ith decision maker is voting "yes," and we don't know what everyone else is voting; and $\phi(0_i, \mathbf{x}) \equiv (x_1, \ldots, x_{i-1}, 0, x_{i+1}, \ldots, x_n)$, which means that the ith decision maker is voting "no," and we don't know what everyone else is voting.

** That is, $\phi(1_i, \mathbf{x}) \geq \phi(0_i, \mathbf{x})$, for all combinations of $\mathbf{x}$; if the ith person decides to vote "yes," the group is not driven towards voting "no."

*** $\mathbf{y} < \mathbf{x}$ means $y_i \leq x_i$ with $y_i < x_i$ for some i. $\mathbf{y} \leq \mathbf{x}$ means $y_i \leq x_i$ for all i.

**** Notation: $\mathbf{a} << \mathbf{b} <=> a_i < b_i$ for all i. Each component of vector $\mathbf{a}$ is strictly less than the corresponding component of vector $\mathbf{b}$.

NOTES

1. That is, instead of voting for option 1 or option 2, individuals vote for option 1 or against option 1; then for option 2 or against option 2. Thus, we are analyzing sequential "yes-no" votes.

2. Competence of the ith decision maker is the probability of the ith individual making the correct choice, i.e.,

$$p_i = p \ \forall \ i$$

3. $\mathrm{Sgn}(x) = x/|x|$; $\mathrm{Sgn}(x)$ equals $+1$ if $x > 0$, -1 if $x < 0$, and is defined as 0 if $x = 0$.

4. Occasionally, we shall refer to $x_i = 1$ as *decisive*, and $x_i = 0$ as *indecisive*, depending on the decision-making context.
5. For n decision-makers, the series operator is: $\sqcap_i x_i \equiv x_1 x_2 \ldots x_n = \min(x_1, \ldots, x_n)$. For n decision makers, the parallel operator is: $\sqcup_i x_i \equiv 1 - \sqcap_i (1 - x_i) = \max(x_1, \ldots, x_n)$.
6. This notation indicates the following:

$\phi(1_i, \mathbf{x}) \equiv (x_1, \ldots, x_{i-1}, 1, x_{i+1}, \ldots, x_n)$, which means that the ith decision maker is voting "yes," and we don't know what everyone else is voting; and

$\phi(0_i, \mathbf{x}) \equiv (x_1, \ldots, x_{i-1}, 0, x_{i+1}, \ldots, x_n)$, which means that the ith decision maker is voting "no," and we do not know what everyone else is voting.

7. That is, $\phi(1_i, \mathbf{x}) \geq \phi(0_i, \mathbf{x})$, for all combinations of $\mathbf{x}$; if the ith person decides to vote "yes," the group is not driven toward voting "no."
8. Consider: $\{1,2,3\}$ $\{3,4,5\}$ cannot be represented by R; $\{1,2,3\}$ $\{4,5\}$ cannot be represented by R; $\{1,3,5\}$ $\{2,3,4\}$ $\{1,2\}$ $\{4,5\}$ (a so-called *bridge* structure) cannot be represented by R; and so on.

REFERENCES

Arrow, K. J. 1951/1963. *Social Choice and Individual Values*, 2nd ed. New York: John Wiley and Sons, Inc.

Banzhaf, J. F. III. 1965. "Weighted Voting Doesn't Work: A Mathematical Analysis." *Rutgers Law Review* 19:317–343.

Barlow, R. E. and F. Proschan. 1981. *Statistical Theory of Reliability and Life Testing: Probability Models.* Silver Spring, MD: To Begin With.

Batchelder, W. H. and A. K. Romney. 1986. "The Statistical Analysis of a General Condorcet Model for Dichotomous Choice Situations." In *Information Pooling and Group Decision Making*, edited by B. Grofman and G. Owen. Greenwich, CT: JAI Press.

Berkowitz, S. D. 1988. "Afterword: Toward a Formal Structural Sociology." In *Social Structures: A Network Approach*, edited by B. Wellman and S. D. Berkowitz. New York: Cambridge University Press.

Bernoulli, J. 1713. *Ars Conjectandi.* Basil: Thurnisiorum.

Birnbaum, Z. W., J. D. Esary, and S. C. Saunders. 1961. "Multicomponent Systems and Structures, and Their Reliability." *Technometrics* 3:55–77.

Birnbaum, Z. W. 1969. "On the Importance of Different Components in Multi-component Systems." In *Multivariate Analysis—II*, edited by P. R. Krishnaiah. New York: Academic Press.

Bower, J. L. 1965a. "Group Decision Making: A Report of an Experimental Study." *Behavioral Science* 10:277–289.

———. 1965b. "The Role of Conflict in Economic Decision Making Groups." *Quarterly Journal of Economics* 70:253–277.

Burt, R. S. 1982. *Toward a Structural Theory of Action: Network Models of Social Structure, Perception, and Action.* New York: Academic Press.

Caplow, T. 1956. "A Theory of Coalitions in the Triad." *American Sociological Review* 21:489–493.

Cartwright, D. and F. Harary. 1977. "A Graph-theoretic Approach to the Investigation of System-environment Relationships." *Journal of Mathematical Sociology* 5:87–111

Casti, J. L. 1989. *Alternate Realities: Mathematical Models of Nature and Man.* New York: John Wiley and Sons, Inc.

Coleman, J. S. 1990. *Foundations of Social Theory.* Cambridge, MA: Harvard University Press.

de Condorcet, N. C. 1785. *Essai Sur L'application de L'analyse a la Probabilité de décisions rendues à Pluralité des Voix.* Paris: Courcier.

De Morgan, A. 1847. *Formal Logic.* Oxford: Clarendon Press.

De Swaan, A. 1973. *Coalition Theories and Cabinet Formations: A Study of Formal Theories of Coalition Formation Applied to Nine European Parliaments after 1918.* San Francisco: Jossey-Bass.

———. 1985. "Coalition Theory and Multi-party Systems: Formal-empirical Theory and Formalizing Approach to Politics." In *Coalition Formation,* edited by H. A. M. Wilke. Amsterdam: North-Holland.

Dion, K. L., R. S. Baron, and N. Miller. 1978. "Why Do Groups Make Riskier Decisions than Individuals?" In *Group Processes: Papers from Advances in Experimental Social Psychology,* edited by L. Berkowitz. New York: Academic Press.

Durant, W. and A. Durant. 1975. *The Age of Napoleon.* New York: Simon and Schuster.

Fararo, T. J. 1989. *The Meaning of General Theoretical Sociology: Tradition and Formalization.* New York: Cambridge University Press.

Foss, R. D. 1981. "Structural Effects in Simulated Jury Decision Making." *Journal of Personality and Social Psychology* 40:1055–1062.

Freeman, L. C. 1989. "Social Networks and the Structure Experiment." In *Research Methods in Social Network Analysis,* edited by L. C. Freeman, D. R. White, and A. K. Romney. Fairfax, VA: George Mason University Press.

French, J. R. P., Jr. 1956. "A Formal Theory of Social Power." *The Psychological Review* 63:181–194.

Gamson, W. A. 1961a. "A Theory of Coalition Formation." *American Sociological Review* 26:373–382.

———. 1961b. "An Experimental Test of a Theory of Coalition Formation." *American Sociological Review* 26:565–573.

———. 1964. "Experimental Studies of Coalition Formation." In *Advances in Experimental Social Psychology,* edited by L. Berkowitz. New York: Academic Press.

Goldberg, S. 1982. *Probability in Social Science: 7 Expository Units Illustrating the Use of Probability Methods and Models, with Exercises, and Bibliographies to Guide Further Reading in the Social Science and Mathematics Literatures.* Stuttgart: Birkhauser Boston, Inc.

Granovetter, M. 1979. "The Theory-gap in Social Network Analysis." In *Perspectives on Social Network Research,* edited by P. W. Holland and S. Leinhardt. New York: Academic Press.

Gray, L. N. and I. Tallman. 1984. "A Satisfaction Balance Model of Decision Making and Choice Behavior." *Social Psychology Quarterly* 47:146–159.

Gray, L. N. and M. C. Stafford. 1988. "On Choice Behavior in Individual and Social Situations." *Social Psychology Quarterly* 51:58–65.

Gray, L. N., M. C. Stafford, and I. Tallman. 1991. "Rewards and Punishments in Complex Human Choices." *Social Psychology Quarterly* 54:318–329.

Grofman, B. 1978. "Judgmental Competence of Individuals and Groups in a Dichotomous Choice Situation: Is a Majority of Heads Better Than One?" *Journal of Mathematical Sociology* 6:47–60.

Grofman, B., G. Owen, and S. L. Feld. 1983. "Thirteen Theorems in Search of the Truth." *Theory and Decision* 15:261–278.

Grofman, B. and S. L. Feld. 1986. "Determining Optimal Weights for Expert Judgement." In *Information Pooling and Group Decision Making,* edited by B. Grofman and G. Owen. Greenwich, CT: JAI Press.

———. 1988. "Rousseau's General Will: A Condorcetian Perspective." *American Political Science Review* 82:567–576.

Grofman, B. and G. Owen. 1982. "A Game Theoretic Approach to Measuring Degree of Centrality in Social Networks." *Social Networks* 4:213–224.

———. 1986. "Condorcet Models, Avenues for Future Research." In *Information Pooling and Group Decision Making,* edited by B. Grofman and G. Owen. Greenwich, CT: JAI Press.

Harnett, D. L. 1967. "A Level of Aspiration Model for Group Decision Making." *Journal of Personality and Social Psychology* 5:58–66.

Hastie, R. 1986. "Experimental Evidence on Group Accuracy." In *Information Pooling and Group Decision Making,* edited by B. Grofman and G. Owen. Greenwich, CT: JAI Press.

Hastie, R., S. D. Penrod, and N. Pennington. 1984. *Inside the Jury*. Cambridge, MA: Harvard University Press.

Herrnstein, R. J. 1970. "On the Law of Effect." *Journal of the Experimental Analysis of Behavior* 17:243–266.

———. 1974. "Formal Properties of the Matching Law." *Journal of the Experimental Analysis of Behavior* 21:159–164.

Holloman, C. R. and H. W. Hendrick. 1972. "Adequacy of Groups Decisions as a Function of the Decision-making Process." *Academy of Management Journal* 15:175–184.

Karotkin, D., S. Nitzan, and J. Paroush. 1988. "The Essential Ranking of Decision Rules in Small Panels of Experts." *Theory and Decision* 24:253–268.

Kerr, N. L., R. Atkin, G. Stasser, D. Meek, R. Holt, and J. Davis. 1976. "Guilt Beyond a Reasonable Doubt: Effects of Concept Definition and Assigned Decision Rule on the Judgements of Mock Jurors." *Journal of Personality and Social Psychology* 34:282–294.

Keynes, J. M. 1921. *A Treatise on Probability*. London: Macmillan.

Komorita, S. S. 1974. "A Weighted Probability Model of Coalition Formation." *Psychological Review* 81:242–256.

Komorita, S. S. and D. A. Kravitz. 1983. "Coalition Formation." In *Small Groups and Social Interaction*, Volume 2, edited by H. H. Blumberg, A. P. Hare, V. Kent, and M. Davies. New York: John Wiley and Sons, Inc.

Kravitz, D. A. 1987. "Size of Smallest Coalition as a Source of Power in Coalition Bargaining." *European Journal of Social Psychology* 17:1–21.

Kushner, H. W. and G. De Maio. 1977. "Using Digraphs to Determine the Crucial Actors in a Voting Body." *Sociometry* 40:361–369.

Lewin, K. 1951. *Field Theory in Social Science*. New York: Harper and Brothers.

Mann, I. and L. S. Shapley. 1964. "The a priori Voting Strength of the Electoral College." In *Game Theory and Related Approaches to Social Behavior: Selections*, edited by M. Shubik. New York: John Wiley and Sons, Inc.

Miller, C. E. 1985. "Group Decision Making Under Majority and Unanimity Decision Rules." *Social Psychology Quarterly* 48:51–61.

———. 1989. "The Social Psychological Effects of Group Decision Rules." In *Psychology of Group Influence*, 2nd ed., edited by P. B. Paulus. Hillsdale, NJ: Lawrence Erlbaum Associates.

Mills, T. 1953. "Power Relations in Three-person Groups." *American Sociological Review* 18:351–357.

Moscovici, S. and M. Zavalloni. 1969. "The Group as a Polarizer of Attitudes." *Journal of Personality and Social Psychology* 12:125–135.

———. 1985. "Social Influence and Conformity." In *Handbook of Social Psychology*, Volume II, edited by G. Lindzey and E. Aronson. New York: Random House Publishers.

Nitzan, S. and J. Paroush. 1982. "Optimal Decision Rules in Uncertain Dichotomous Choice Situations." *International Economic Review* 23:289–297.

———. 1984a. "The Significance of Independent Decisions in Uncertain Dichotomous Choice Situations." *Theory and Decision* 17:47–60.

———. 1984b. "A General Theorem and Eight Corollaries in Search of Correct Decision." *Theory and Decision* 17:211–220.

———. 1984c. "Partial Information on Decisional Competencies and the Desirability of the Expert Rule in Uncertain Dichotomous Choice Situations." *Theory and Decision* 17:275–286.

Raiffa, H. 1968. *Decision Analysis: Introductory Lectures on Choices under Uncertainty*. Reading, MA: Addison-Wesley.

Russell, B. 1948. *Human Knowledge: Its Scope and Limits*. New York: Simon and Schuster.

Savage, L. J. 1954. *The Foundations of Statistics*. New York: Wiley.

Shapley, L. S. and B. Grofman. 1984. "Optimizing Group Judgmental Accuracy in the Presence of Interdependencies." *Public Choice* 43:329–343.

Shapley, L. S. and M. Shubik. 1954. "A Method for Evaluating the Distribution of Power in a Committee System." *American Political Science Review* 48:787–792.

Shubik, M. 1982. *Game Theory in the Social Sciences: Concepts and Solutions*. Cambridge, MA: The MIT Press.

Simmel, G. 1950. *The Sociology of Georg Simmel*. Translated, edited and with an introduction by K. H. Wolff. New York: The Free Press.

Tallman, I. and L. N. Gray. 1990. "Choices, Decisions, and Problem-solving." *Annual Review of Sociology* 16:405–433.

Thompson, L. L., E. A. Mannix, and M. H. Bazerman. 1988. "Group Negotiation: Effects of Decision Rule, Agenda, and Aspiration." *Journal of Personality and Social Psychology* 54:86–95.

Vinokur, A. 1971. "A Review and Theoretical Analysis of the Effect of Group Processes upon Individual and Group Decisions Involving Risk." *Psychological Bulletin* 76:231–250.

Von Bertalanffy, L. 1968. "General Systems Theory—A Critical Review." In *Modern Systems Research for the Behavioral Scientist*, edited by W. Buckley. Chicago: Aldine Publishing Company.

Von Neumann, J. and O. Morgenstern. 1944. *The Theory of Games and Economic Behavior*. Princeton, NJ: Princeton University Press.

Wang, M., K. W. Hipel, and N. M. Fraser. 1988. "Modeling Misperceptions in Games." *Behavioral Science* 33:207–223.

Watson, S. R. and D. M. Buede. 1987. *Decision Synthesis: The Principles and Practice of Decision Analysis*. New York: Cambridge University Press.

Wellman, B. 1988. "Structural Analysis: From Method and Metaphor to Theory and Substance." In *Social Structures: A Network Approach*, edited by B. Wellman and S. D. Berkowitz. New York: Cambridge University Press.

Young, H. P. 1988. "Condorcet's Theory of Voting." *American Political Science Review* 82:1231–1244.

Zellner, A. 1971. *An Introduction to Bayesian Inferences in Econometrics*. New York: John Wiley and Sons, Inc.

STRUCTURE, CULTURE, AND INTERACTION:
COMPARING TWO GENERATIVE THEORIES

Cecilia Ridgeway and Lynn Smith-Lovin

ABSTRACT

We compare affect control theory and expectation states theory to assess their usefulness in developing micro-macro linkages. Both theories suggest that people use cultural meanings to translate larger social structures into interpersonal interaction. These cultural meanings generate patterns of behavior, but also are affected by behavioral patterns. This feature allows the theories to predict the conditions under which structures will be maintained or changed. Each theory's strengths compliment the other's weaker areas: Expectation states is more precise in its translation of macro features into cultural knowledge, while affect control theory is more precise about how interaction can affect cultural meaning.

Recently, some theorists of macrosociological processes have called for a better understanding of the way systems of human action mediate macro-level outcomes.

Advances in Group Processes, Volume 11, pages 213–239.
ISBN: 1-55938-857-9

DiMaggio and Powell (1991), for instance, noted that a theory of individual action was necessary to support the macrocultural processes featured in their new institutionalization approach. As Coleman (1986) has pointed out, many sociological models that link macro-level causes to macro-level outcomes implicitly assume some degree of actor-level enactment or mediation of the process. But the precise nature of this mediation is rarely specified. The result, at a minimum, is a limitation in our ability to understand the actual causal processes by which certain important sociological outcomes such as institutionalization or stratification occur.[1] Even when macro-level outcomes can be *predicted* from macro-level antecedents, micro-level theorizing is often necessary to *explain* the intervening causal processes by which the predicted relationship occurs (Coleman 1986). For those macro-level outcomes where actors not only mediate, but introduce change through actor-level processes, the loss is not only in explanation but also in adequate prediction of the outcomes themselves.

The challenge is to rectify this situation by locating or developing social psychological theories that specify precise connections between clearly defined features of macrosociological systems and patterns of interaction among actors. In our view, a variety of specific theories will be necessary. We focus here on two theories that propose particular links between some features of macro-level systems and systems of action: affect control theory (Heise 1978, 1979; Smith-Lovin and Heise 1988) and expectation states theory (Berger, Conner, and Fisek 1974; Webster and Foschi 1988).

Despite many differences, these two theories are distinctive among theories of interaction in that they take a generative (as we explain shortly) and situational approach to explaining social action. These aspects, if further developed, could be useful in the construction of macro-micro-macro accounts. Yet, to our knowledge, these two theories have never been systematically compared. We know little of their underlying similarities or differences. In fact, the theories have rather different origins. Expectation states theory developed from research in the Bales tradition of small group studies (Berger et al. 1974; Meeker 1981). Affect control theory derived from research on the measurement of cultural meanings and from psychological research on impression formation (Heise 1978, 1979).

The purpose of this paper is to examine the relationship between the theories with an eye toward assessing their potential to provide accounts of the actor mediation of important macro-level processes. By noting similarities between the theories, we hope to point to theoretical features that may be necessary (or at least useful) in establishing micro-macro connections. By comparing the theories and examining their relationship, we also highlight weaknesses that must be addressed and possible points of mutual reinforcement.

The key feature shared by affect control and expectation states theories that may prove useful for micro-macro theorizing is that they are both generative theories of behavior rather than script-oriented theories. Both theories seek to show how macro-level features of the social environment constrain interaction in such a way as to create microstructures of action that enact and support those macro-level features. In this they are like many other sociological approaches to social psychology. What is distinctive is how they account for this micro enactment of macro-level features. The theories eschew the common approach of viewing such interaction as the enactment of prescribed rules, roles, or social scripts. Instead, they view actors first as possessing basic information on the cultural meaning of socially significant categories by which persons, settings, and events can be classified. Second, actors are presumed to possess rules for combining the information that is evoked by a given situation in order to generate a resultant course of action toward others and self in that situation. This combination of basic information and rules for combining it to address the contingencies of particular situations is what we refer to as a generative theory of behavior.

To see the implications of generative versus script approaches to behavior, it is useful to consider the analogy of language. A foreign-speaking tourist can take two approaches to dealing with the new language. She can learn by rote a list of phrases or sentences common to basic interactions such as "Pleased to meet you, my name is. . . or "Where is the post office?" This will be useful, but soon she is likely to find herself in a situation that varies significantly from the exact sentences she knows. She is stymied because she knows no way of varying or rearranging the words she knows to respond to the unexpected situation. She would be better off if she took the second approach of learning basic words and some grammatical rules for combining them. Then she could generate new sentences to respond to new situations. It is the fact that language is a generative system that allows speakers to construct an infinite number of sentences that are meaningful and intelligible to others from a finite vocabulary and a few rules of grammar (Chomsky 1978). Because affect control and expectation states theories take a similarly generative approach to behavior, they depict interactants as responding flexibly to contingent events while still enacting and maintaining macrocultural (i.e., widely shared) meanings.

Why should generative theories of behavior be especially useful for constructing theories of the actor mediation of macro-level processes? Given that a particular macro-level process is actor mediated, this mediation could take two forms. Events at the actor level could be wholly determined by macro-level constraints so that the micro events are a mere conduit for macro-level causal processes. This is the type of micro mediation that script-oriented theories handle very well. However, in this type of mediation, a micro analysis adds only explanatory, not predictive power to a purely macro-level analysis that derives outcomes directly from the initial macro-level constraints.

In the second form of mediation, micro events may introduce changes in the macro-level processes in the process of mediating them. Fararo and Skvoretz (1986), for instance, note that for a network shape or formal hierarchy to be stable, interactional outcomes among actors must reflect their position in the network or hierarchy. Yet process effects in interaction mean that hierarchical outcomes are not always fully predictable from actors' initial resources and attributes (Chase 1980). This gives interactional patterns a dynamic quality that allows them, through the outcomes they create, to have a causal impact on the larger structures in which they are embedded. In this form of mediation, where the addition of a micro-level account adds predictive as well as explanatory power, the limits of script-oriented approaches can be seen. Innovative or unexpected interactional outcomes that undermine a structural pattern are, almost by definition, not those that go according to script.

Whether the interactional mediation of macro-level processes is fully determined or has an independent causal impact is likely to vary with the particular phenomenon of interest. Resolution of the issue and the relative weights of macro and micro causes in a particular case are essentially empirical questions. But what matters here is that both forms of mediation occur. To be most useful in the construction of micro-macro accounts, a theory of interactional mediation must be able to manage both. Generative theories of behavior do this. Like script-oriented theories, they can show how fully determined patterns of interaction are carried out. But unlike script theories, they can also show how responses to unpredictable contingencies of events can, under some circumstances, introduce change into the macro-level process. We argue that affect control theory and expectations states theory allow both types of mediation, and therefore have special potential for exploring multi-level links.

We provide a very brief sketch of each theory's structure for those who are not familiar with them. Then, we consider several questions about the relationship between affect control theory and expectations states theory. What are the features of macro-level systems whose impact on interaction each theory explicates? We assess the strengths and weaknesses of these structural conceptions and make suggestions for broadening them. Next we discuss the scopes of interaction to which the theories apply and the implications of these scope restrictions for macro-micro theorizing. We then review the processes posited by each theory. What are the theory's views of the dynamic processes of interaction? What is the nature of the feedback to the macrosystem envisioned by the theory? What do their views of dynamic processes in interaction suggest about the role of interaction in maintaining and changing macro-structural patterns? We end with suggestions for future development, both within the theories themselves and for other theories that are developing that attempt to link macro-structures to micro-interaction.

A BRIEF SKETCH OF THE THEORIES

Affect Control Theory

Affect control theory is, at its heart, a *control* system. Social action is generated by the reference signals that are created by an actor's definition of the situation.

Interaction is interpreted in light of those reference signals, and new action is expected, intended and (in the absence of physical or structural constraints), enacted to maintain that definition. Thus, interaction serves to control our perception of the situation in which we are engaged. The behaviors that we generate serve to maintain our view of ourselves and others within the situation. When events occur that disturb that view, we generate new action to restore our original definition.

To make specific statements about the interrelationship of social perception and social behavior as a control system, affect control theory must translate both elements into a common metric—a conceptual substrate that allows social identities, behaviors, and other event elements to be related to one another in a systematic way. Affect control theory accomplishes this goal by using three dimensions of affective meaning—evaluation, potency, and activity—that psycholinguistic research demonstrated were fundamental to widely varied conceptual domains and language communities (Osgood, Suci, and Tannenbaum 1957; Osgood, May, and Miron 1975). The evaluation dimension contrasts good vs. bad, niceness vs. nastiness. The potency dimension contrasts the powerful with the weak, the deep with the shallow. The activity dimension contrasts that which is lively and expressive with that which is quiet, reserved, or withdrawn.

When people enter a social situation, they develop a cognitive view of themselves and others that represents the institutional context of interaction, their goals when entering the situation, and/or their relationships to other actors in the setting. The social identities that express this cognitive definition of the situation carry affective meanings on the evaluation, potency, and activity dimensions. People also view potential social actions (cognitively labeled behaviors) with these same dimensions; acts can be nice or nasty, powerful or weak, lively or quiet. These affective meanings are the reference signals (called fundamental sentiments) that are controlled and maintained by interaction.

Of course, events that happen during social interaction can change the meanings that we hold for ourselves, other people, and the behaviors that are enacted. Using the common evaluation-potency-activity metric, affect control researchers developed equations that describe how events can change the meanings of identities and behaviors within social interaction (Smith-Lovin 1987). The reaction equations combined with the control assumption allow the theory to predict how meanings are related dynamically to social action. These equations are the theory's "grammatical rules" specifying how people recombine the affective meanings culturally associated with the identities and events in the situation to generate social action.

Normally, actions are generated that, when perceived and processed as events, maintain meanings. What actions will maintain these meanings, of course, depend on how we see ourselves (our social identity in the situation and its meanings for us), who we are acting upon, and how we view them (their identities and the corresponding meanings). So mothers act differently toward fathers and children, and expect different actions from them. This differentiation occurs because fathers are much more powerful but less lively than children (both are quite nice). A woman would act differently in her mother identity than in her professional identity, both because these identities have different meanings and because she would be acting toward different objects.

But, of course, things happen in interaction that can cause changes in meanings within the situation. People can define the situation differently, can hold different meanings for the same identities or behaviors, or can accidently produce behaviors that are misinterpreted by another. When these occasions occur, actions that maintain one actor's meanings are disturbing for another's view of the situation (see a more lengthy discussion below in the section on micro-macro feedback). These disturbing events lead to new action to restore meanings (Robinson and Smith-Lovin 1992). Affect control theory predicts that new behaviors will be expected and enacted that, when processed, will cause reactions that move meanings back in line with the reference signals (the original view of the situation and its meanings). In the control system, micro-level actions are generated by the cultural meanings, but the micro-level interaction is also the mechanism through which these meanings are maintained (or, under unusual circumstances, altered).

Expectations States Theory

Expectation states is a theory of status relations among people who interact. It explains how, when people share a collective task or goal, they organize themselves, often inadvertently, into observable hierarchies of power and prestige. Like affect control theory, expectations states theory assumes that, to interact, people need definitions of self and others in the situation. The definitions they develop, in turn, control their behavior. When the situation is task-oriented, defining self and others in relation to task success or failure is the crucial component of this process that controls each actor's behavior in regard to the task effort and creates the hierarchy. The theory conceptualizes these self-other definitions as performance expectations, which are anticipations of the likely usefulness of one actor's contributions to the shared task compared to each other's. Actors usually hold performance expectations as implicit, out of awareness assumptions about self and others rather than as conscious judgments.

If an actor forms a more positive expectation for self than for another, the actor is more likely to speak up, offer opinions about the task, and stick to them if challenged. On the other hand, if the actor's expectation for self is less positive than for the other, the actor is more likely to hesitate in making suggestions, to ask for and respond positively to the other's opinions, and to change opinions to agree with the other. Thus, the theory asserts that the differential expectations held for self compared to all others cause actors to behave in a manner that creates corresponding inequalities in the opportunities they receive to participate, how much they do participate, the evaluations their contributions receive from others, and their influence over collective decisions. In this fashion, actors' efforts to work together on a collective task result in an observable hierarchy of power and prestige.

Actors create performance expectations by drawing upon consensual cultural beliefs about the evaluative and competence implications of actors' attributes and behaviors in the situation. The consensual belief systems addressed by expectations states theory are those that associate socially recognized attributes and behavioral styles of actors with status value (i.e., social worthiness and competence) or the possession of socially valued rewards. The theory defines several different status value belief systems. Widely held beliefs about status characteristics associate membership in various social categories (e.g., sex, ethnic, occupational categories) with greater or lesser status value, competence, and often specific skill levels as well (Berger et al. 1977). Status typification beliefs link actively assertive vs. deferential behavior with high status-low status, leader-follower relations (Fisek, Berger, and Norman 1991). Formalized status elements are organizational positions or honors associated in cultural beliefs with greater or lesser status value (e.g., dean vs student; chair vs member; prize winner vs others). The theory also specifies referential belief systems that associate attributes of actors with the possession of greater or lesser reward levels (e.g., the belief that men typically earn more than women) (Berger, Fisek, Norman, and Wagner 1985). As this indicates, the macro-cultural (i.e, consensual) belief systems discussed by expectation states theory are largely cultural representations (or cultural assertions) of evaluative and material hierarchies in society.

Actors possess such consensual beliefs as part of their cultural vocabulary. But the theory asserts that these beliefs only affect actors' performance expectations and behavior in a given situation when the beliefs become salient in that situation. In their efforts to define self and other in order to act, actors notice attributes (e.g., race, gender, age, dress, job title, or education) or behaviors (e.g., speech style, accent, or a confident or nervous manner,) that either differentiate self and others in socially evaluated ways or that are culturally defined as relevant to the task. When they notice such an attribute or behavior, it makes salient for them the widely shared cultural beliefs about it.

Beliefs carrying both positive and negative status associations, activated by the specific contingencies of the situation, are then combined to form aggregate expectations for self and others that become the basis of situational behavior. This is what makes expectation states a generative theory. The cultural beliefs form the vocabulary and the rules of combining and aggregating are the grammar that produces situationally specific expectations that drive behavior.

The theory posits a formally specified set of rules by which people weigh activated status beliefs by their relevance to the situational goals and combine their positive and negative implications to form aggregate performance expectations for each person compared to the others (Berger, Fisek, Norman, and Zelditch 1977). These formal combining rules allow the theory to make precise predictions about the relative power and prestige people with diverse attributes and behaviors will have in a given setting (Balkwell 1991). It can predict, for instance, how influential a young African-American lawyer will be interacting with middle-aged European-American salesmen.

The theory assumes that everybody follows the same rules in combining information to form expectations. Consequently, assuming actors in a situation share similar consensual beliefs and have the same information about each other's distinguishing attributes and behaviors, they will form equivalent self-other expectations. Because of this, performance expectations tend to become self-fulfilling in interaction. Because both the physician and the mechanic on the jury unthinkingly assume the physician is more competent, the physician argues forcefully and the mechanic defers regardless of either's actual knowledge of the case at hand. The self-fulfilling effects of performance expectations on interaction are the means by which interactional hierarchies come to reflect and maintain the structures of inequality in macro-level systems.

REPRESENTATIONS OF STRUCTURE

Culture, Structure, and Interaction

As these overviews reveal, both theories, despite their differences, assume that the structural features of society are translated into interactional settings through widely shared cultural beliefs and meanings. In other words, the features of macro-level systems whose impact on interaction are explicated are cultural (involving shared beliefs), rather than directly structural (involving material patterns of relations). However, these cultural beliefs are taken for granted representations of different aspects of the macro-structural context. It may be the use of cultural representations of macro-structure that allows these theories to be generative rather than script-oriented. Macro-cultural (i.e., consensual) beliefs provide actors with a flexible and adaptable interface between the concrete constraints of macro-structure and the contingencies of micro-level interaction.

Sewell (1992), reformulating Giddens (1976, 1984), provides an account of culture/structure that can help us explicate the role of cultural representations in affect control and expectation states theory, showing its relationship to their generative approaches to action and the reproduction of structure. Sewell defines structure in a manner that incorporates both culture and structure as we use the terms. Structure is both schemas and resources, he claims. He notes that there are different levels of such structures in social relations, some on the surface and apparent and some "deeper" and more taken for granted. It is clear that the cultural belief systems that constitute the "vocabulary" of both theories (i.e, fundamental sentiments about identities and consensual status value beliefs) are schemas in Sewell's sense that represent macro-level material arrangements in society. Thus, both our theories access macro-level structures most directly through the schema half of the schema-resources duality that constitutes structure by Sewell's account.

Interestingly, however, in both theories the "grammatical rules" for combining these cultural vocabularies of schemas are also themselves schemas of a sort. But they represent much deeper "structures," not of interactional relations, but of the social/cognitive processes of impression formation (affect control theory) or the functional evaluation of information in task oriented situations (expectation states theory). These theories are generative precisely because they predict action from the intersection of two levels of structures (in Sewell's sense). The demands of one structuring process (e.g., the formation of impressions or aggregated performance expectations) come together with the structuring implications of schemas from macro-level schema/resource systems.

The two levels of structure intersect because they are both evoked by events at a third level, that of the interactional situation. The joint evocation of schemas from macro-level structures and deeper social/cognitive schemas provide the means by which actors create emergent structures that organize their interaction. Since such interactional microstructures are the product of more than one structuring process, they can create elements that would not be produced by any one structuring process alone. As a result, they have a theoretically specified potential to introduce (or block the introduction of) situational schemas (e.g., situational identities or self-other status relations) that challenge some of the macro-level schemas on which they are based. These theories predict behavior in a generative fashion, then, because they provide an account of the intersection of multiple schema/resource structures at multiple levels.

Macro-Structural Representations

Despite their homologous use of consensual cultural schemas to link macro and micro levels, our two theories differ in the nature of the cultural beliefs they rely upon and the elements of macro structure represented in these beliefs. Expectation states theory essentially addresses the impact of widely shared beliefs about the stratification of the macro-level system within which the actors function. As everyday representations of the evaluative and material structure of inequality surrounding the context of interaction, these beliefs have relatively specific denotative as well as connotative, affective meanings. This is different from the more inclusive, summative, and strictly connotative cultural meanings addressed by affect control theory.

In affect control theory, positions within the social structure are represented as identities with varying degrees of status (esteem), power, and expressivity. These are specified by the identities' location in the three dimensional space representing the terms' affective meaning. These cultural meanings often are communicated directly during the socialization process (e.g., through emotional expressions or interaction patterns). However, their original development occurs through the control of material resources and meaningful ritual action.

For example, we may learn that police are powerful people by seeing our parents' fleeting expression of anxiety when approaching a policeman or through their deferential manner during the subsequent interaction. But the meaning associated with the position resulted from the policemen's control over force (weapons and the trained ability to use them) and from the institutional routines that give police the right to arrest and charge suspects with crimes. Status or esteem, translated into an evaluative reaction of goodness or badness, results from the positional resources that allow one to produce pleasurable, beneficial results (Demerath 1993). Power, translated into the potency dimension of affective meaning, results from control over physical force or valued resources. Expressivity, translated into the activity dimension, reflects both the technical activity associated with the position and the degree of engagement or withdrawal with other actors within the institutional framework.

Translating an enormously variable set of social positions into a three-dimensional space of cultural meaning provides additional benefits. We can see the interactional similarity between two positions that have roughly the same position within the affective meaning space, even when these positions have rather different technical activities. For example, the identities "secretary," "housewife," and "sister-in-law" all have roughly similar affective profiles (somewhat nice, slightly powerless, and somewhat lively—about 1.0, -0.5 and 1.0 on evaluation, potency, and activity scales that range from +4 to -4). While these positions vary from office to household to family institutions, the affective cultural meaning represents the interactional similarity of the roles dictated by their lively pleasantness combined with lack of control over important resources. Similarly, mothers and nurses are quite close affectively—both have the goodness, moderate potency, and liveliness necessary to produce nurturant, caring activities toward those who are less potent (children or patients, respectively), while being deferent to institutional participants with more power (fathers and doctors).

Affect control theory emphasizes people's use of cultural meanings to create a composite interpersonal gestalt (Heise 1979, pp. 3–8). Space-time coordinates of action, the setting's material organization, the goals of the participants, and the institutional context of the action are all important in developing a cognitively sensible whole. As a result, affect control theory is more likely than expectation states theory to deal with formal, institutionalized identities that are evoked by a specific organizational setting. Indeed, the macro-level process that could be linked most obviously to affect control theory's actor-mediated account is institutionalization (Smith-Lovin and Douglass 1992).

In spite of these general principles, however, the link between macro-structure and cultural meaning is much less clearly specified in affect control theory than in expectation states theory. Expectation states theory better explicates the impact of macro-level, trans-organizational structures of inequality such as those based on wealth, class, occupational prestige, education, gender, or race, on a specific level of micro-level outcomes, power and prestige in interaction. Because the theory links macro-level inequality with structures of micro-level inequality, the macro-level process it can be linked to most obviously is stratification. Indeed, the systematic incorporation of an actor-level account such as expectation states theory may be necessary to achieve an understanding of the role of status in macro systems of stratification (Lawler, Ridgeway, and Markovsky 1993).

The reliance of both theories on the definition of the situation through cultural representations of structure gives them a thoroughly sociological cast and an ability to introduce innovation into larger social structures. However, it makes both vulnerable to a potential criticism. In effect, both theories posit that a difference or an advantage must be cognitively recognized to affect the outcomes of interaction. Statuses must become salient to lead to expectations about competence in expectations states theory; identities must be adopted and behaviors recognized to create the affective reactions that drive behavior in affect control theory. Are both theories subject, then, to the common criticisms of symbolic interactionism: that it ignores the importance of material structure? The gun that is not noticed can still kill, such critics say.

Both theories survive such attack through their ability to incorporate unexpected events into new definitions at the interaction level. Structural features that are not initially incorporated into one's definition of the situation will affect subsequent interaction, both because they will enable certain physical activities (i.e., acquiring goods, providing information, or causing harm) and because they may lead inter-actants to define the situation in opposing ways (e.g., the young upstart who knows that she possesses expertise in the area of discussion). Both expectations states and affect control theories provide mechanisms through which actions or private knowledge can influence interaction and lead toward a new consensually held definition. In expectation states theory, material structures affect its dynamics through the impact of resources on assessments of worth and implied ability. These in turn encourage actions that modify self-other expectations. In affect control theory, the behaviors that such structures enable allow the labeling of actors with new identities as people ask "what kind of a person could do such a thing?"

SCOPE OF THE THEORIES

The scope and domain of expectation states and affect control theories differ substantially. While affect control theory applies to virtually any co-oriented interaction, expectation states theory restricts its scope to interaction that focuses on a collective task. The "task," however, can be interpreted broadly to include any shared goal for which the actors have a sense of success or failure, from a committee task or work project to a family buying a house or choosing a movie to see. The domain of expectation states theory is also more restricted. That is, within task-oriented interaction, the theory only addresses the task directed behaviors that make up the observable power and prestige order (i.e., participation, attention and evaluation, and influence). The theory acknowledges that even in the most task-oriented setting other processes such as affective relations and social control also go on. It also recognizes that these processes may occasionally interact with task oriented processes to moderate or magnify the power and prestige structure (Berger 1988). But the theory analytically abstracts out the task processes that drive the formation of the hierarchy and explicates these alone. As a result, the theory provides much more specific analyses of how interactional hierarchies form and change than does affect control theory.

Affect control theory offers a less detailed but more inclusive representation of the multidimensional total of interaction. It includes an explicit treatment of how dynamics involving esteem, power, and expressivity interrelate during a social interaction. For example, the impression change processes that describe actors' reactions to events (Smith-Lovin 1987) show that power use toward weak alters make one lose the esteem of observers, at the same time that one's potency is enhanced.

The place of emotions within the two theories illustrates the trade-off between generality and specificity of prediction. In expectations states theory, Ridgeway and Johnson (1990) have produced predictions about the prevalence of positive and negative socioemotional responses in high and low status actors within a task setting. High status actors are expected to produce both praise and criticism, while low status actors produce only praise for the (assumed) higher quality productions of their status superiors. The socioemotional dynamics apply only to the task domain; the positive and negative affect are, in effect, evaluations of task perform-ances. Affect control theory, on the other hand, predicts a constant stream of emotional experience as the result of both identity maintenance and situations that fail to maintain identities. In stable, identity maintaining situations, the affective character of the identity determines the typical affective state; when disturbing events occur, emotions are determined both by the identities' meaning and by the degree and direction of deflection created by the event (Smith-Lovin and Heise 1988; Smith-Lovin 1990).

The difference in domain results in another difference between the theories. Expectation states theory acknowledges that performance expectations are only one aspect of the situated identities actors define for self and others in interaction. However, performance expectations, as an aspect of situated identity, cannot be easily represented as one of three dimensions by which affect control describes summative identities. This creates a difficulty in relating the two theories.

Clearly, identities that are higher on evaluation and potency in affect control theory generally receive higher performance expectations than identities that are lower on evaluation and potency. Therefore, affect control theory has the closest connection to the status characteristics branch of expectations states theory (Berger, Wagner, and Zelditch 1985), where general identities (like gender and social class) that are activated by differentiation within the immediate setting cause inferences about competence (Smith-Lovin and Robinson 1991). In affect control theory, these eifects are modeled by modifying institutional identities with status characteristics (e.g., a "judge" becomes a "female judge" when the identity schema indicates that it is usually occupied by a male, and that the female occupant therefore deserves a linguistic marker).

Task-specific competencies are less easily handled by affect control theory. When the discussion turns to sewing, the female status becomes the one that is most tied to the task. In expectations states theory, the female would be accorded a higher expectation for competence (although this prediction is based on our knowledge of the culture, not on the formal theoretical structure). In affect control theory, the female interactant would gain a higher evaluation and potency only by attaining the identity of "expert" within the context of the discussion. Thus, while performance expectations are a part of the composite identities addressed by affect control theory, they cannot be easily disaggregated from the composite and expressed in affect control theory terms. There is no simple translation between performance expectations and fundamental sentiments, despite the homologous role these concepts play within their respective theories.

Despite these differences of scope and domain, there are remarkable similarities between the structure of these theories as generative theories of structure and action. Both evoke consensual beliefs systems as the mediators between macro-level structures and micro-level action. Both theories argue that people must develop a definition of an interaction situation before a coherent pattern of behavior is generated. Both theories make rather strong assumptions about the consensual nature of cultural beliefs that guide interaction, and the fact that typical interactions tend to maintain and reaffirm these cultural beliefs. The theories differ in the scope of the dynamics that they encompass. As we shall see, they also differ in the degree of elaboration with which they treat different stages of the dynamic process of interaction.

DYNAMICS OF INTERACTION

Definition of the Situation

Both theories suggest that people's cognitive processing of a situation is key to predicting the structure of the ensuing interaction. This is the mechanism through which structural/cultural contexts are imported to interactional settings. The theories differ somewhat in their view of the definition process.

Expectation states theory formally specifies the conditions under which particular cultural beliefs will be activated and used to form the set of performance expectations for each actor that constitute the definition of the situation addressed by the theory. According to the theory, people notice attributes or behaviors that differentiate them or that are culturally defined as relevant to the task. The process is primarily perceptual and does not necessarily involve conscious awareness.

The emphasis in affect control theory is somewhat different. The definition process is explicitly cognitive, with interactants actively involved in creating an interpersonal gestalt that defines sensible relations for various pairs of actors (Heise 1979, pp. 3–8). Beyond that, affect control theory has been much less specific than expectation states theory about *how* situations are defined (although Heise 1979 gives some ideas), and much more specific about how behaviors are generated *after* the definition is obtained.

The two theories also differ in the attention they have given to the way interactants define one another in situations where they have multiplex relationships. This is the question of how people combine identities when they have two or more salient relations toward one another in a single situation. In contrast to expectations states theory's extensive treatment of the status combination process, affect control theory has only a preliminary view of how such relationships are handled.

In an institutionalized context, where one of the salient identities is clearly primary (i.e., "physician" in a hospital), affect control theory can predict how the institutionalized identity will be modified by status characteristics. Averett and Heise (1987) showed how the meanings of identities shift when they are modified by characteristics like male, female, young, old, competent, incompetent, and so forth. The process here is similar to that in expectations states theory, in that differentiation within the group spurs cognitive recognition of such qualifiers (e.g., a female physician in a group of male physicians is more likely to have her gender noticed). Also like expectation states theory, the less task relevant qualifier modifies but does not overwhelm the task relevant institutional identity. However, affect control theory gives more emphasis to the overall institutional schema that underlies the identity. "Female physician" would be a more likely modification than "male physician," even when both are participating in a sex-differentiated group, because the cognitive schema for "physician" includes the information that the identity is usually occupied by a male. Therefore, the modifier "female" might be activated even in a group of physicians that was all female.[2]

However, when an institutional context does not define a single salient identity as primary, multiplex relationships are more difficult for affect control theory to handle. In current versions of the affect control theory simulation program INTER-ACT (Heise 1991), multiple identities can be averaged to form composites. Therefore, someone who is both a husband and a colleague in a given situation would enact an affective meaning that is an even mix of the two. Another possibility is that people engage in parallel processing, assessing the implications of events for each identity that they hold in relation to the other; this could be the source of mixed emotions experienced in response to events that support one identity but deflect meanings associated with another (Smith-Lovin 1992). Clearly, affect control theory needs more work on this issue to bring its predictions to the level of specificity produced by expectations states theory.

Affect control theory also has not developed a detailed treatment of situations that are initially undefined. The model is capable of working from observed behaviors to labeling of interactants by asking, "what kind of person would act in this way" and solving for the identity profile. But affect control theorists would consider an undefined interaction situation as rare and extremely transient. More likely is a situation in which behaviors (or emotion displays) are used to disambiguate a situation and to choose between a small number of institutionally relevant possibilities (Robinson, Smith-Lovin, and Tsoudis 1994; Smith-Lovin 1990).

In contrast, the interaction of status equals (and the subsequent development of power and prestige orders within the group) is a classic, much studied case for expectation state researchers (Berger et al. 1974; Fisek, Berger, and Norman 1991). Like affect control, expectation states theory argues that undefined situations are clarified through initial behaviors that then evoke defining cultural beliefs that create performance expectations. However, expectation states theory specifies this process in much greater detail than affect control theory and relies less on surrounding institutional cues.

According to the theory, when actors are undifferentiated in expectations (i.e., a homogeneous group or a homogeneous subset of a group), if one actor for any reason initiates a higher status behavior (task contribution, evaluation, influence, confident demeanor) than another, then there is an above chance probability that the next behaviors involving the two actors will have the same status implications (Fisek et al. 1991). Thus status behaviors, once initiated, tend to lead to similarly high or low status behaviors and fall into a pattern where A acts high status and B acts low status. At some point, group members implicitly notice these behavior patterns, which makes salient cultural status typifications of leader-follower people and causes the members to form differentiated expectations for A and B. Once A has been elevated above B in the group, there is an increased likelihood that other group members not already A's superior will also treat A as higher status and that members not already B's inferior will treat B as lower status. Thus the hierarchy of performance expectations and behavior tends to spread throughout the group and become defined in a transitive fashion. In this way, undefined situations are defined.

Production of Behavior

In our introduction, we mention two types of mediation, structurally induced mediation in which interaction enacts larger cultural or structural dictates and mediation with micro innovations that don't fully reproduce larger forms. Both affect control theory and expectation states theories posit that the cultural information evoked by a situation generates behaviors that tend to support that initial assessment. Consequently, both theories clearly emphasize structurally induced mediation. In affect control theory, this is the behavioral confirmation of fundamental sentiments while in expectation states theory it is the generalization of external status characteristics to power and prestige in the situation.

However, both theories also emphasize that structurally induced patterns are an interactional accomplishment in the face of multiple contingent events that must be managed to maintain the structure. Furthermore, both theories agree that the initial definition of the situation can draw upon complex sets of cultural beliefs and combine them in ways for which no well defined behavioral expectation exists (e.g., in multiplex relationships). Finally, both theories agree that behavioral events can, under some circumstances, cause actors to redefine the situation. Consequently, in both theories, the same processes for the production of behavior result in enactments that reproduce macro-level arrangements under many circumstances, but more innovative and idiosyncratic interactional structures under other circumstances.

Although both agree that behavior is kept consistent with situational definitions, the two theories differ slightly in the motivational means by which this occurs. In affect control theory, the maintenance of situational meaning is seen as the controlling mechanism behind behavior. Thus the theory posits a rather strong collective motivation on the part of actors to behave in ways that confirm the definition of the situation. When disconfirming events occur, they elicit predictable affective reactions, according to the theory.

In expectation states theory, on the other hand, initial performance expectations generate confirming behaviors in a self-fulfilling process that is less affectively motivated and almost accidental. Such is the presumed power of this self-fulfilling process that it is only more recently that the theory has seriously addressed what happens when expectations are disconfirmed. Under specified conditions, when behavior challenges a legitimated power and prestige order, the theory predicts affectively motivated behavioral reactions to maintain expectations, much as affect control theory does (Ridgeway and Berger 1986). When legitimacy is not an issue, however, the theory predicts that disconfirming events will be treated more neutrally as new information (e.g., Wagner, Ford, and Ford 1986). A review of each theory's account of the production of behavior clarifies these and other differences.

Expectation states theory describes a powerful self-fulfilling process by which the set of aggregate performance expectations that actors hold for each compared to the others (i.e., their initial definition of the situation) generate task behaviors that create a confirming behavioral hierarchy. Only when the quality of contributions by group members is clearly perceived to contradict the pre-existing order of performance expectations would the interaction fail to reinforce and support this structure. But the theory notes a number of mechanisms by which this self-fulfilling process is achieved in the face of recurrent contradictory events.

There are several potential sources of disconfirming behaviors in any goal-oriented interaction. Cultural beliefs activated in the situation by the actors' status characteristics or other attributes are imperfectly correlated with their actual differences in confidence, task skills, interaction skills, and motivation, all of which affect the quality of their performances. In addition, behavioral performances must be brought off in interaction. Some element of pure contingency can lead to a surprising behavioral success or failure (i.e., people can blow it or luck out). Also, another actor's performance or reaction can inadvertently cast your performance in an unexpectedly good or bad light. Finally, contingent events occasionally produce salient outside "tests" (i.e., the boat sinks) or other evaluations of performances.

Expectation states theory posits a number of information processing biases induced by performance expectations that reduce the likelihood that such disconfirming performances will be perceived as such. Expectations bias the attention an actor receives so that good contributions from a low status actor may simply not be noticed by others or credited to that actor. Expectations also bias evaluations, so that the same suggestion "sounds better" coming from a high status (i.e., high performance expectations) actor than from a low status actor. Finally, double standards for inferring ability mean that even if an actor's performance is acknowledged to be good (or clearly bad), it will be taken as a stronger evidence of ability (or weaker evidence of inability) for a high status actor than a low status one (Foschi 1991). All these biases make it harder for disconfirming events to interrupt the self-fulling process and help keep interaction in line with the larger structure.

Biases, however, can not prevent all contradictory events from being recognized. Once a given round of behaviors from two actors is interpreted as contradicting their expected status relation, it increases the likelihood that another such behavioral exchange will occur, leading to a contradictory behavior pattern. How the group reacts to this unexpected behavior will depend on the group context.

When the power and prestige order challenged by the behavior has not (or not yet) acquired any special legitimacy in the eyes of its members or when there is outside support for the new behavior, members will revise their performance expectations to accommodate the new behavior (Ridgeway and Berger 1986; Wagner, Ford, and Ford 1986). In other words, they will redefine the situation and change the power and prestige structure. According to the theory, the unexpected behavior pattern activates contradictory status typifications (cultural beliefs about high and low status behaviors), replacing prior behavioral typifications for the two actors and modifying aggregate expectations for them. The reevaluation of the two actors spreads change through the hierarchy, affecting in varying degrees all who are not securely superior or inferior to both in expectations and behavior (Fisek et al. 1991).

If, on the other hand, the existing status order of the group is legitimate in the eyes of the actors, then the theory predicts that a contradictory behavior pattern will encounter resistance (Ridgeway and Berger 1986). Orders are most likely to acquire legitimacy when they are consistently supported by consensual status beliefs from a larger social entity or by performance evaluations by an outside authority. Evidence shows, for instance, that simple behavioral assertiveness from a member with low status characteristics often evokes negative reactions and fails to acquire influence (Carli 1990; Butler and Geis 1990).

So legitimacy processes moderate the extent to which behavior inconsistent with expectations is allowed to change a status order that supports consensual cultural beliefs or authorized evaluations. These are the interactional status orders that are most clearly determined by the larger stratification system. Like affect control theory, expectation states theory interprets the legitimacy effects that bolster status orders as a meaning maintenance process by which actors keep things as they "should be."

The interactional accomplishment of legitimacy processes, in turn, give status orders that support cultural patterns greater resistance to contingent events than more idiosyncratic status orders. By demonstrating how this occurs, expectation states theory illustrates the complex interactional processes by which face-to-face status relations are maintained in predominant accord with a society's macro-level structure of inequality. By showing how structure-maintaining processes, while powerful, are not inevitable (they can be deflected at several points), the theory also indicates how interactional status orders can undermine existing structures of inequality. It is the tension between the faithful enactment of cultural beliefs and the unpredictable and not fully controllable contingencies of interaction that create the dynamics of status processes.

We see again, then, that within its scope and domain, expectation states theory generates a fairly concrete and specific account of the way the definition of the situation produces behavior that usually confirms, but sometimes redefines, the situation. Affect control theory, in contrast, has a much wider scope and domain of application. In its predictions about how behaviors are produced to maintain meanings imported from the larger structural/cultural context, we again see how this greater generality creates both power and ambiguity.

In affect control, the predicted quantity is not a specific pattern of behaviors (e.g., A will talk more than B) but an affective profile for the behavior that, when processed, will cause the least deflection and therefore maintain fundamental meanings. Actual behaviors are predicted by searching a corpus of labeled actions and their affective meanings to find a good match. So, for example, a mother might want to behave in a very nice, somewhat potent manner toward a son (the actual prediction is 2.28, 1.14 and −0.19 respectively on the +4 to −4 evaluation, potency and activity scales). Such a profile matches behaviors like "compliment," "approve of," and "talk to." These labels are a cognitive representation of what she will intend to do, and what an alter who shares her definition of the situation will expect of her. But what will constitute a compliment in the context of this relationship is not specified. Behaviors are open to definition and interpretation, just like actors. So we see that affect control theory is again broader, more inclusive, but less concrete than expectations states theory.

The two theories' views of the production of behavior also provide the clearest distinction between their predictions. In expectations states theory, the behaviors that actors produce are dependent only on the *relative* expectations about their task contributions, with little or no input from their absolute position. If A thinks she is better than B and C at the current task and they agree with this assessment, the fact that A is below average in competence in some wider population has little importance in the immediate situation. A will direct activity, talk more, evaluate the contributions of others, and so on. Only the relative differences salient in the situation (however they are arrived at) determine behavior.

In affect control theory, on the other hand, both absolute position within the three-dimensional space *and* relative position determine the character of behavior. For example, a mother is roughly the same distance from a son on the evaluation, potency, and activity dimensions as a policeman is from a callgirl; in both cases, the latter is roughly 0.8 units lower in evaluation (on a scale that runs from +4 to −4), 1.1 units lower in potency, and 1.7 units higher in activity than the former. But a mother is predicted to compliment a son (an ideal behavior profile of 2.3, 1.1 and −0.2 on the three dimensions respectively), while the policeman is predicted to interview the callgirl (0.7, 1.1, 0.0). The difference is created by the fact that the mother has a more positive (high evaluation) identity than the policeman, and is interacting with a more positive object-person. Relative differences do matter in affect control theory: the mother would caution a spectator and the policeman would reprimand a pothead (again, a comparison that maintains the *relative* differences between actor and object in the two interactions). Both actors get more negative when the differences between their evaluation and their object persons' evaluation get larger. But their absolute position (and that of their object) has a definite impact on the behavior expected. In expectations states theory, it is *relative* status only that produces the behavioral predictions.

When resources do not allow culture-maintaining enactments (e.g., we are high status, but we do not know a thing about the topic at hand) or when random variations occur in behavioral routines, both theories allow these creative interaction patterns to feed back into the beliefs of interaction partners. The dynamic aspect of affect control theory is especially clear in its predictions once such disturbing behaviors occur. Under the theory, the actions that will *restore* meanings are *not* simply a reenactment of the behaviors that would have maintained the original identities if no disturbing event had occurred. The interaction partners and behaviors will be chosen for their value in restoring the fundamental sentiments. For example, Wiggins and Heise (1987) found that students who had been criticized by a secretary engaged in very positive behaviors to restore their damaged self esteem if they were with another student, but behaved somewhat negatively if they were interacting with a delinquent.[3] Affect control theory predicted this pattern, since the reaction equations contain a "balance" effect: doing good unto a good object produces upward movement in evaluation, as does doing something negative to a bad object.

We have seen how the definitions of self and others influence behavior in affect control theory (through setting the reference signals that generate and guide interpretation of behavior). We also have seen how events in interaction lead to changes in behavior patterns for identity occupants, as interactants attempt to restore meanings after disturbances. The final process through which the dynamics of interaction feed back into the initial situation is through potential change in the definition of the situation. If events are consistently disturbing in a particular direction (e.g., a person acts consistently more powerfully than we expect), the interaction partner eventually will change his or her cognitive view of that person's identity. The original identity may be modified with a qualifying characteristic (unusually *mature* child), or the identity itself may be changed (when someone you thought was a beginner turns out to be an expert). A more subtle, long-term change may occur when fundamental meanings shift over time as a result when events consistently deflect meanings in one direction (e.g., when recurrent scandals lower our evaluation of the identity "politician"). All of these processes—modification, reidentification, and meaning change—produce long-term changes in behavioral patterns within a situation (or even within a culture), since they change the reference signals that interactants are attempting to maintain during interaction (see the discussion of micro-macro links following for a fuller discussion of these issues).

THE NATURE OF FEEDBACK TO MACRO SYSTEM FROM MICRO INTERACTION

Both theories predict that interactional situations can be redefined in ways that erode rather than maintain established cultural beliefs and the structural arrangements they represent. Thus, while both emphasize the maintenance of structural/cultural forms, both can account for interactionally generated change in those forms. Despite the differences we noted in each theory's account of interactional dynamics, the two agree substantially about the conditions under which these dynamics will feed change back into the macro-level.

In both theories, the primary thrust of such change will be in the consensual belief systems that provide the interface between structure at the macro-level and the organization of micro-level interaction. In expectation states theory, consensual beliefs about the status value of individual attributes and behaviors are the cultural forms that link the macro-structure of inequality with the micro-dynamics of face-to-face status relations. It is these status value beliefs that interactionally generated change is most likely to modify.

On the other hand, for affect control theory, the change will be in the affective meanings (fundamental sentiments) by which identities are defined. It is the identity of an individual within an interactional situation that represents how that individual fits into larger organizational contexts and the network of relationships that constitute social structure. Since the concept of identity links macrostructure and microinteraction, interactionally generated changes in the cultural meanings of identities will modify structural forms, according to affect control theory.

To predict interactionally generated change in affect control theory, we need to look to situations that will produce consistent deflections of meaning. To predict changes in status value beliefs, expectation states theory directs us towards situations that consistently produce aggregate performance expectations for actors that do not support basic cultural patterns about the kinds of attributes and behaviors that indicate worthiness and competence. The theories agree upon three types of situations that will consistently produce such meaning deflecting or structure—undermining situational definitions for their participants, eventually modifying consensual beliefs.

First (and perhaps least interesting from a microinteractional point of view), are occasions when the material resources that support a behavioral repertoire are changed exogenously. If the actions that would support an identity's meaning are not physically possible (e.g., the policemen cannot control the suspects, because the suspects now have 50mm machine guns), then restorative events cannot be created. Similarly, if people advantaged by a status characteristic no longer control the resources to appear more worthy and able than those disadvantaged by the characteristic, then the status value of the characteristic becomes difficult to maintain. This might happen, for instance, when an occupation such as typesetter is deskilled, making it difficult for typesetters to act more worthy and able than typists.

In such situations, redefinition or meaning change will occur after the new behavioral repertoire becomes established. Changes like this occur when system-wide changes in size, technology, or resources limit or enable different action patterns. Such shifts can be very important; the reason they are relatively uninteresting to us here is that the microinteractional change is merely a conduit for one structural change (in material circumstances) to be translated into another (organizational and cultural forms).

The second process leading to macro-change is one of diffusion created when differing cultures or subcultures meet.[4] When people define a situation differently, or hold differing meanings for the same identity, or associate different status value beliefs with the same attribute, their interaction will produce disturbing events for one or more interactants. Events that are confirmatory for one participant are deflecting or disconfirming for the other(s). If the interactional contact between the differing "world views" is common or enduring, cultural change is likely to occur. For example, this may be the process by which beliefs about the status value of distinctive behaviors or possessions (e.g., a Japanese or American car) spread among subcultures to society at large.

In an example from affect control theory, research has shown that when two people from different cultures marry, they bring to that union somewhat different images of what husbands and wives are like. Japanese culture expects husbands to be much more powerful but also much quieter than their wives; the United States views both roles as relatively potent and lively (Okiyama, cited in Smith-Lovin 1988). If Japanese-American marriages became much more common after World War II, one would expect the images of these familial roles to shift, especially in the populations where the intermarriages were most likely to occur (e.g., military families). Several features of this situation are relevant: (1) the interaction is long-term and consistent, (2) at least some behavioral outputs are relatively unambiguous and cannot be altered through maintenance-biased cognitive processing, and (3) the contact between the cultures is likely to be repeated across a large number of individuals. Again, however, this case of diffusion through cultural contact is another instance of macrostructure (networks) operating through microinteraction to produce cultural change.

The third case is potentially the most interesting, since the micro-level response is key to the macro change. This is the case where multiplex relationships lead to changes in the meaning of identities or in beliefs about their relative status value. Affect control theory would describe the innovative effects of multiplexity as follows.

While one identity may be central during a given multiplex interaction, other identities that are activated by the relationship doubtless will be processed simultaneously. If the multiple identities are similar in character, then actions that maintain one will cause little disturbance in the other(s). However, if one person must simultaneously maintain very disparate identities/meanings in a single interaction, then deflection of situated meanings from fundamental sentiments is bound to occur. If women are seen as very nice but not too powerful, and judges are seen as neutral on evaluation but extremely powerful, then it will be difficult for a female judge to interact with male colleagues in the criminal justice system (not to mention male perpetrators) without either being too nice for a judge or too powerful for a woman. If such a combination becomes frequent throughout the system, the meanings of judge and/or woman must change. The key things about this change is that after it occurs throughout the system, it will change the ways in which women are viewed in other contexts that have nothing to do with their employment. The fundamental meaning of the "woman" identity will have shifted, and maintaining this new meaning will make women operate differently toward other object-persons whenever the woman identity is evoked by the situation. (Perhaps judges will be nicer people, too, throughout the system.)

In this way, demographic change in the correlation of identities is translated through microinteraction into new meanings at the cultural level. It is the mixed emotions that individuals feel while trying to maintain disparate identities that lead to intraindividual meaning change. It is the multiple layered perception of the interaction (I see this person as both a judge and a woman) that leads to change in alters' meanings. Demographic change in identity occupancy will have the most effect on cultural meanings when identity performances cannot be physically isolated from one another (e.g., by keeping work and home lives separate) and when the meanings of conjoint identities are maximally different (e.g., one occupies a nice identity at the same time as a deviant one).

Expectation states theory offers a similar argument about the way structural or demographic changes in the correlation among status valued attributes and material resources in a population can lead to reoccurring, multiplex interactions that alter a given status value belief or even create a new one (Ridgeway 1991). Such structural changes increase the number of interactions where people have status characteristics that are inconsistent with each other or with the actor's skills and reward levels (i.e., a highly paid female expert interacts with a poorly paid non-expert male). The aggregate performance expectations people form in these situations result in those who have a low state of an important status characteristic (e. g., a female in mixed sex interaction) being thought more competent and worthy in the interaction than those with a high state of the status characteristic (a male). This "reversal" of expectations is only in the specific situation. However, research has shown that such reversal experiences transfer to the next person of that status characteristic (e.g,. female or male) that the actor deals with, although with diminishing effect (Pugh and Wahrman 1983; Markovsky, Berger, and Smith 1984). This creates a diffusion process through which reversal experiences spread out and begin to modify the expectations of the wider population.

Overtime, the belief-modifying effects of such structurally induced reversal experiences begin to accumulate, gradually shifting the content of the status value beliefs that define the status characteristics. In this example, for instance, the status value of being male rather than female would be reduced. Once the consensual status value belief is modified, it changes performance expectations formed for men and women (and their consequent power and prestige) in all task oriented interaction, not just in reversal situations. In this way, the structurally induced but interactionally generated change in the status value belief spreads change in the structure of inequality throughout the society (Ridgeway 1991). These processes can increase or reduce the status significance of an individual attribute that already carries status value in society (e.g., gender) or associate new status significance with an attribute that previously carried little status value (Ridgeway 1991).

This discussion makes evident an important feature of the micro-macro link in both theories here. While chance, innovation, tacit knowledge, and so on, may influence local meanings within a given interaction, such deviations must occur *systematically* across a larger number of interactions to have an effect on larger culture. One competent, forceful woman eventually may change her male colleagues opinions about her competence; such interactions must be repeated widely and often in the occupational structure to change the cultural view that women in general are less capable than men. Since culture (i.e., shared belief systems) recreates itself through interaction unless it is definitively and repeatedly contradicted, it has a chameleon-like ability to regenerate inequalities in the face of dramatically shifting structural circumstances (Ridgeway 1993). Only structural changes that create consistent, necessary, repeated reflections in micro-interaction will cumulate in cultural change that in turn propagates and maintains new structural arrangements (Ridgeway 1991).

CONCLUSIONS

We began by suggesting that generative theories of micro-interaction could be useful in establishing micro-mediation of macro-level changes. Both affect control theory and expectations states theory emphasize the degree to which people actively generate structure in local interaction, rather than simply enacting a prescribed set of behaviors. The agency with which both theories invest actors allows them to explain both the surprising continuity of macro-forms, as structure is imported into interactive situations and the potential for change (both local and system-wide). Note that these theories achieve the flexibility introduced by a generative model without losing the predictive nature of a theoretical framework. Actors in these theories have agency—they produce behaviors to reproduce their understandings of the situation and respond creatively to new situations that disturb that view. But they do not act in a haphazard or capricious way. In both theories, a "vocabulary" of cultural meanings is transformed by a set of processual rules in a manner that can be modeled and understood by social scientists. While the output of this system is as variable as the situations in which actors find themselves, the process is quite regular. It maintains the macro structure in most cases, but can produce innovative outcomes at the system level when micro-interactions are disturbed in the same manner over a large part of the system.

In both cases, the theories have focused on cultural meaning as the mechanism through which material structures are translated into micro-interaction. It is the meanings (status expectations or affective characterizations) that individual actors are socialized to hold through their past interactions; it is these meanings that they carry into situations and use to generate actions. Other generative theories focus on more material aspects of structure (Skvoretz 1984; Skvoretz and Fararo 1980; Axten and Skvoretz 1980; Heise 1988a, 1988b), but even they depend heavily on the cultural information embodied in institutionalized arrangements of resources and actors. We suspect that culture is a key link between macro-structural features and the micro-level interaction that supports (and in some cases, changes) them.

Given these substantial similarities, we note some variations in the strengths and weaknesses of expectation states theory and affect control theory. Expectation states theory has a much more elaborated description of how macro features are translated into the cultural knowledge that it uses to make predictions (i.e., performance expectations). It also has a somewhat more specific treatment of the process through which actors develop a definition of a situation (i.e., the expectations hierarchy). Given its scope and domain, the macro-level process that expectation states theory's micro-account could most directly be linked to is stratification.

Affect control theory is wider in scope and deals with a much broader range of situations and behaviors. It is more explicit about how the definition of the situation generates behaviors, and the interaction of the dimensions of cultural meaning. Affect control's strengths lie in the prediction of emotional and behavioral responses to disturbing events, and to the reinterpretation that occurs when that disturbance builds up to an intolerable level. With its focus on situated but culturally formed identity, affect control theory may provide a useful account of actor mediation in the macro-level process of institutionalization.

Since the theories differ so much in scope and domain, we cannot envision any crucial test that would support one at the expense of the other. More likely, they will both prove useful (as will other generative theories of micro-interaction) in creating the interactional substrate that enacts, supports, and occasionally changes the structural context of interaction. On one point, however, the theories do seem to offer somewhat differing views. Expectations states theory uses only relative status information to predict behavior, while affect control theory gives weight to both absolute position and relative position. In the domains where their predictions overlap, this apparent contradiction may prove useful in comparing the two.

ACKNOWLEDGMENTS

The authors would like to acknowledge the support of National Science Foundation grants SES 9210171 to Cecilia Ridgeway and SES 9008951 to Lynn Smith-Lovin. We thank Linda Molm, Lisa Rashotte and two reviewers for comments on an earlier version of the paper.

NOTES

1. Indeed, Heise (1975) argues that we could not establish a causal relationship definitively without specifying an *operator* that translated the variation of the causal variable into the variation of the effect variable.
2. Note the similarity to Ridgeway's (1988) discussion of status characteristics' activation by the larger organizational context, rather than the composition of the immediate group. The difference between the two is subtle: in affect control theory, it is cultural information (the identity schema) that leads to the linguistic marker; in Ridgeway's argument, it is the larger organization in which the group is embedded.
3. The secretary and student (or delinquent) actually were confederates.
4. Such diffusion processes are examples of what Sewell (1992) refers to as the transposition of schemas across differing situations as structures interact.

REFERENCES

Averett, C. P. and D. R. Heise. 1987. "Modified Social Identities: Amalgamations, Attributions, and Emotions." *Journal of Mathematical Sociology* 13:103–132.

Axten, N. and J. Skvoretz. 1980. "Roles and Role-Programs." *Quality and Quantity* 14:547–83.

Balkwell, J. 1991. "Status Characteristics and Social Interaction: An Assessment of Theoretical Variants." Pp. 135–176 in *Advances in Group Processes*, Vol. 8, edited by E. Lawler, B. Markovsky, C. Ridgeway, and H. Walker. Greenwich, CT: JAI Press.

Berger, J. 1988. "Directions in Expectations States Theory." Pp. 450–476 in *Status Generalization: New Theory and Research*, edited by M. Webster and M. Foschi. Stanford, CA: Stanford University Press.

Berger, J., T. L. Conner, and M. H. Fisek, eds. 1974. *Expectation States Theory: A Theoretical Research Program*. Cambridge, MA: Winthrop.

Berger, J., M. H. Fisek, R. Z. Norman, and M. Zelditch Jr. 1977. *Status Characteristics and Social Interaction: An Expectation States Approach*. New York: Elsevier Scientific.

Berger, J., M. H. Fisek, R. Z. Norman, and D. G. Wagner. 1985. "The Formation of Reward Expectations in Status Situations." Pp. 215–226 in *Status, Rewards, and Influence: How Expectations Organize Behavior*, edited by J. Berger and M. Zelditch, Jr. San Francisco: Jossey-Bass.

Berger, J., D. G. Wagner, and M. Zelditch, Jr. 1985. "Introduction: Expectations States Theory: Review and Assessment." In *Status, Rewards and Influence*, edited by J. D. Berger and M. Zelditch, Jr. San Francisco: Jossey-Bass.

Butler, D. and F. L. Geis. 1990. "Nonverbal Affect Responses to Male and Female Leaders: Implications for Leadership Evaluations." *Journal of Personality and Social Psychology* 58:48–59.

Carli, L. L. 1990. "Gender, Language, and Influence." *Journal of Personality and Social Psychology* 59:941–51.

Chase, I. 1980. "Social Process and Hierarchy Formation in Small Groups." *American Sociological Review* 45:905–24.

Coleman, J. 1986. "Social Theory, Social Research, and a Theory of Action." *American Journal of Sociology* 91:1309–1335.

Chomsky, N. 1978. *Syntactic Structures*. The Hague: Mouton.

Demerath, L. 1993. "Knowledge-based Affect: Cognitive Origins of "Good" and "Bad."" *Social Psychology Quarterly* 56:136–147.

Dimaggio, P. J. and W. W. Powell. 1991. "Introduction." Pp. 1–38 in *The New Institutionalism in Organizational Analysis*, edited by W. W. Powell and P. J. Dimaggio. Chicago: University of Chicago Press.

Fararo, T. J. and J. Skvoretz. 1986. "E-State Structuralism." *American Sociological Review* 51:591–602.

Fisek, M. H., J. Berger, and R. Z. Norman. 1991. "Participation in Heterogeneous and Homogeneous Groups: A Theoretical Integration." *American Journal of Sociology* 97:114–142.

Foschi, M. 1991. "Gender and Double Standards for Competence." In *Gender, Interaction, and Inequality*, edited by C. Ridgeway. New York: Springer-Verlag.

Giddens, A. 1976. *New Rules of Sociological Method: A Positive Critique of Interpretive Sociology*. London: Hutchinson.

———. 1984. *The Constitution of Society: Outline of the Theory of Structuralism*. Los Angeles: University of California Press.

Heise, D. R. 1975. *Causal Analysis*. New York: Wiley.

———. 1978. "Behavior as the Control of Affect." *Behavioral Science* 22:163–177.

———. 1979. *Understanding Events: Affect and the Construction of Social Action*. New York: Cambridge University Press.

———. 1988a. "Computer Analyses of Cultural Structures." *Social Science Computer Review* 6:183–196.

———. 1988b. "Modeling Event Structures." *Journal of Mathematical Sociology* 13:138–68.

———. 1991. *Interact 2: A Computer Program for Studying Cultural Meanings and Social Interactions*. Department of Sociology, Indiana University.

Lawler, E., C. Ridgeway, and B. Markovsky. 1993. "Structural Social Psychology and the Micro-Macro Problem." *Sociological Theory* 11:268–290.

Markovsky, B., L. F. Smith, and J. Berger. 1984. "Do Status Interventions Persist?" *American Sociological Review* 49:373–82.

Meeker, B. F. 1981. "Expectation States and Interpersonal Behavior." Pp. 290–319 in *Social Psychology: Sociological Perspectives*, edited by M. Rosenberg and R. Turner. New York: Basic Books.

Osgood, C. E., G. C. Suci, and P. H. Tannenbaum. 1957. *The Measurement of Meaning*. Urbana: University of Illinois Press.

Osgood, C. E., W. H. May, and M. S. Miron. 1975. *Cross-cultural Universals of Affective Meaning*. Urbana: University of Illinois Press.

Pugh, M. D. and R. Wahrman. 1983. "Neutralizing Sexism in Mixed Sex Groups: Do Women Have to be Better Than Men?" *American Journal of Sociology* 88:746–63.

Ridgeway, C. L. 1988. "Gender Differences in Task Groups: A Status and Legitimacy Account." Pp. 188–206 in *Status Generalization*, edited by M. Webster and M. Foschi. Stanford, CA: Stanford University Press.

————. 1991. "The Social Construction of Status Value: Gender and Other Nominal Characteristics." *Social Forces* 70:367–86.

————. 1993. *"Face-to-face With Gender."* Paper presented at the annual meetings of the American Sociological Association, Miami, Florida.

Ridgeway, C. L. and J. Berger. 1986. "Expectations, Legitimation and Dominance Behavior in Task Groups." *American Sociological Review* 51:603–17.

Ridgeway, C. and C. Johnson. 1990. "What is the Relationship between Socioemotional Behavior and Status in Task Groups?" *American Journal of Sociology* 95:1189–1212.

Robinson, D. T. and L. Smith-Lovin. 1992. "Selective Interaction as a Strategy for Identity Maintenance: An Affect Control Model." *Social Psychology Quarterly* 55:1:12–28.

Robinson, D. T., L. Smith-Lovin, and O. Tsoudis. 1994. "Emotion Displays and Sentencing in Vignettes of Criminal Trials." *Social Forces* 73(1).

Sewell, W., Jr. 1992. "A Theory of Structure: Duality, Agency, and Transformation." *American Journal of Sociology* 98:1–29.

Skvoretz, J. 1984. "Languages and Grammars of Action and Interaction: Some Further Results." *Behavioral Science* 29:81–97.

Skvoretz, J. and T. J. Fararo. 1980. "Languages and Grammars of Action and Interaction: A Contribution to the Formal Theory of Action." *Behavioral Science* 25:9–22.

Smith-Lovin, L. 1987. "Impressions from Events." *Journal of Mathematical Sociology* 13:35–70.

————. 1988. "Affect Control Theory: An Assessment." Pp. 171–192 in *Analyzing Social Interaction: Advances in Affect Control Theory*, edited by L. Smith-Lovin and D. R. Heise. New York: Gordon and Breach Science Publishers.

————. 1990. "Emotion as Confirmation and Disconfirmation of Identity: An Affect Control Model." Pp. 238–70 in *Research Agendas in Emotions*, edited by T. D. Kemper. New York: SUNY Press.

————. 1992. "An Affect Control View of Cognition and Emotion." In *Self and Society: A Social Cognition Approach*, edited by J. Howard and P. Callero. Cambridge: Cambridge University Press.

Smith-Lovin, L. and W. T. Douglass. 1992. "An Affect Control Analysis of Two Religious Subcultures." In *Social Perspectives on Emotions*, edited by D. Franks and V. Gecas. Greenwich, CT: JAI Press.

Smith-Lovin, L. and D. R. Heise. 1988. *Analyzing Social Interaction: Advances in Affect Control Theory*. New York: Gordon and Breach Science Publishers.

Smith-Lovin, L. and D. T. Robinson. 1991. "Gender and Conversational Dynamics." Pp. 122–156 in *Gender and Interaction: The Role of Microstructures in Inequality*, edited by C. Ridgeway. New York: Springer-Verlag.

Wagner, D. G., R. S. Ford, and T. W. Ford. 1986. "Can Gender Inequalities Be Reduced?" *American Sociological Review* 51:47–61.

Webster, M. A. Jr. and M. Foschi. 1988. *Status Generalization: New Theory and Research.* Stanford, CA: Stanford University Press.

Wiggins, B. A. and D. R. Heise. 1987. "Expectations, Intentions and Behavior: Some Tests of Affect Control Theory." *Journal of Mathematical Sociology* 13:153–69.

Advances in Group Processes

J A I   P R E S S

Edited by **Edward J. Lawler,** *Department of Organizational Behavior, Cornell University*

REVIEWS: A major impression one gets from this volume is that far from being dormant, the social psychology of groups and interpersonal relations is quite vibrant, and very much involved with compelling problems.

Concerns about the imminent demise of group processes as an area of study in social psychology are clearly exaggerated. But should doubts remain, they ought to be allayed by the range and quality of the offerings in this volume.

. . . should be of interest both to specialists in group processes and to sociologists who are interested in theory, particularly in theoretical linkages between micro and macro analysis. Because many of the papers offer thorough reviews and analyses of existing theoretical work, as well as new theoretical ideas, they are also useful readings for graduate students."

-- Contemporary Sociology

Volume 10, I993, 304 pp. $73.25
ISBN I-55938-280-5

CONTENTS: Preface, *Edward J. Lawler, Barry Markovsky, Karen Heimer and Jodi OBrien.* Social Learning and the Structure of Collective Action, *Michael W. Macy.* Cross-National Experimental Investigations of Elementary Theory: Implications for the Generality of the Theory and the Autonomy of Social Structure, *David Willer and Jacek Szmatka.* Politics and Control in Organizations, *Gerald R. Ferris, Julianne F. Brand, Stephen Brand, Kendrith M. Rowland, David C. Gilmore, Thomas R. King, K. Michele Kacmar, and Carol A. Burton.* How Sentiments Organized Interaction, *Robert K. Shelly.* The Influence of Group Processes on Pseudoscientific Belief: "Knowledge Industries" and the Legitimation of Threatened Worldviews, *Raymond A. Eve and Francis B. Harrold.* An Expected Value Model of Social Exchange Outcomes, *Noah E. Friedkin.* Approaching Distributive and Procedural Justice: Are Separate Routes Necessary? *Karen A. Hegtvedt.* Affect and the Perception of Injustice, *Steven J. Scher and David R. Heise.* What is a Group? A Multilevel Analysis, *Per Manson.* Prior Social Ties and Movement into New Social Relationships, *Richard T. Serpe and Sheldon Stryker.*

Also Available:
Volumes 1-10 (1984-1993) $73.25 each

Research in
the Sociology of Organizations

Edited by **Samuel B. Bacharach,** *New York State School of Industrial and Labor Relations,* Cornell University

Associate Editors: **Peter Bamberger,** *Bar Ilan University,* **Pamela Tolbert,** *Cornell University* and **David Torres,** *University of Illinois, Chicago Circle*

REVIEW: Research in the Sociology of Organizations is designed as an annual review of current work related to organizations. The editor apparently identifies individuals with major scholarly programs related to organizations and invites them to summarize their work. All six chapters in this, the second volume, clearly reflect several years of work by the authors. All are intellectually independent; other than emphasizing something about organizations, they have little in common.

-- *Contemporary Sociology*

Volume 12, Special Issue on Labor Relations and Unions
1993, 309 pp $73.25
ISBN 1-55938-736-X

Edited by **Ronald Seeber,** *Cornell University* and **David Walsh,** *Miami University*

CONTENTS: Introduction, *Samuel Bacharach, Ronald Seeber, and David Walsh.* Issues in Union Structure, *George Strauss.* Integrating U.S. Labor Leadership: Union Democracy and the Ascent of Ethnic and Racial Minorities and Women into National Union Office, *Daniel B. Cornfield.* The Formation and Trasformation of National Unions: A Generative Approach to the Evolution of Labor Organizations, *Peter D. Sherer and Huseyin Leblebici.* National Union Effectiveness, *Jack Fiorito, Paul Jarley, and John Thomas Delaney.* Unions and Legitimacy: A Conceptual Refinement, *Gary N. Chaison, Barbara Bigelow, and Edward Ottensmeyer.* Conflict Resolution and Management in Contemporary Work Organizations; Theoretical Perspectives and Empirical Evidence, *David Lewin.* A Diagnostic Approach to Labor Relations in Organizations, *Arie Shirom.* The Labor Movement as an Interorganizational Nework, *David J. Walsh.* Associational Movements and Employment Rights: An Emerging Paradigm?, *Charles Heckscher and David Palmer.*

Also Available:
Volumes 1-11 (1982-1993) $73.25

J A I   P R E S S

J A I   P R E S S

Contemporary Studies in Sociology
Theoretical and Empirical Monographs

Volume 12.

Self, Collective Behavior and Society:
Essays Honoring the Contributions of Ralph H. Turner

Edited by **Gerald M. Platt,** *Department of Sociology, University of Massachusetts* and **Chad Gordon,** *Department of Sociology, Rice University*

1994, 413 pp.   LC 94-3519 $73.25
ISBN: 1-55938-755-6

CONTENTS: Preface, *Neil J. Smelser.* Ralph Turner and His Sociology: An Introduction, *Gerald M. Platt and Chad Gordon.* I. ORIGINS OF A SOCIOLOGICAL APPROACH. The Birth of a Book: Collaboration, Personal and Professional, *Lewis M. Killian.* Influences of Symbolic Interactionism on Disaster Research, *Joanne M. Nigg.* II. SCHOLARLY CONTRIBUTIONS: INTERACTIONIST AND CONSTRUCTIVIST PERSPECTIVES. A. Collective Behavior and Social Movements. Multiple Images of a Charismatic: An Interpretive Conception of Martin Luther King Jr.s Leadership, *Gerald M. Platt and Stephen J. Lilley.* Making Sense of Collective Preoccupations: Lessons from Research on the Iben Browing Earthquake Prediction, *Kathleen J. Tierney.* The Study of Collective Behavior: An Elaboration and Critical Assessment, *David A. Snow and Phillip W. Davis.* B. Self, Modernity and Post-Modernity. Freedom and Constraint in Social and Personal Life: Toward Resolving the Paradox of Self, *Sheldon Stryker.* In Search of the Real Sef: Problems of Authenticity in Modern Times, *Viktor Gecas.* Self, Role, and Discourse, *John P. Hewitt.* In Search of the Post-Modern Self: The Babbitt Brothers in Las Vegas, *Norman K. Denzin.* Self-Systems: Five Current Social Pychological Approaches, *Chad Gordon.* C. Role, Differentiation and Social Change. The Dynamics of Role Enactment, *Wayne E. Baker and Robert R. Faulkner.* Dynamics and Contingencies Instructural Change, *Paul Colomy.* D. Emergence and the Structuring of Social Realities. New Telecommunications Technologies and Emergent Norms, *Gary T. Marx.* The Social Meaning of Meals: Hierarchy, Equality and the American Pot-Luck, *Joseph R. Gusfield.* III. TEACHER AND MENTOR: REMEMBRANCES OF TIMES PAST AND

**J
A
I

P
R
E
S
S**

JAI PRESS INC.
55 Old Post Road # 2 - P.O. Box 1678
Greenwich, Connecticut 06836-1678
Tel: (203) 661- 7602   Fax: (203) 661-0792

Advances in Human Ecology

Edited by **Lee Freese**, *Department of Sociology, Washington State University*

This series publishes theoretical, empirical, and review papers on scientific human ecology. Human ecology is interpreted to include structural and functional changes in human social organization and sociocultural systems as these changes may be affected by, interdependent with, or identical to changes in ecosystemic, evolutionary, or ethological processes, factors, or mechanisms. Three degrees of scope are included in this interpretation: (1) the adaptation of sociocultural forces to bioecological forces: (2) the interactions, or two-way adaptations, between sociocultural and bioecological forces; (3) the integration, or unified interactions, of sociocultural with bioecological forces.

The goal of the series is to promote the growth of human ecology as an interdisciplinary problem solving paradigm. Contributions are solicited without regard for particular theoretical, methodological, or disciplinary orthodoxies, and may range across ecological anthropology, ecological economics, ecological demography, ecological geography, biopolitics, and other relevant fields of specialization.

Volume 3, 1994, 247 pp. $73.25
ISBN 1-55938-760-2

CONTENTS: Preface, *Lee Freese*. The World Around Us and How We Make It: Human Ecology as Human Artifact, *C. Dyke*. The Political Economy of Environmental Problems and Policies: Consciousness, Conflict, and Control Capacity, *Allan Schnaiberg*. The Assembling of Human Populations: Toward a Synthesis of Ecological and Geopolitical Theories, *Jonathan H. Turner*. A Culture Scale Perspective on Human Ecology and Development, *John H. Bodley*. The Ecology of Macrostructure, *Jonathan H. Turner*. Evolutionary Tangles for Sociocultural Systems: Some Clues from Biology, *Lee Freese*. Human Ecology: Lost Grail Found? A Critical Review of Programs to Address Human-Environmental Relationships, *Jeremy Pratt*.

Also Available:
Volumes 1-2 (1992-1993) $73.25 each

JAI PRESS INC.
55 Old Post Road # 2 - P.O. Box 1678
Greenwich, Connecticut 06836-1678
Tel: (203) 661- 7602   Fax: (203) 661-0792